Bernhard Graf

Bridges that Changed the World

Prestel

Munich · Berlin · London · New York

Contents

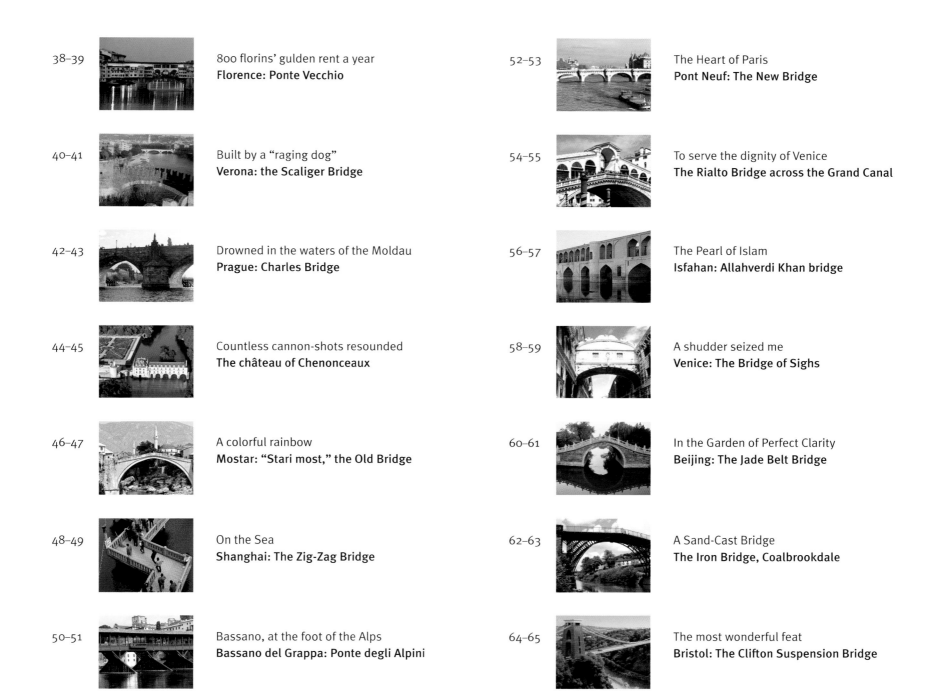

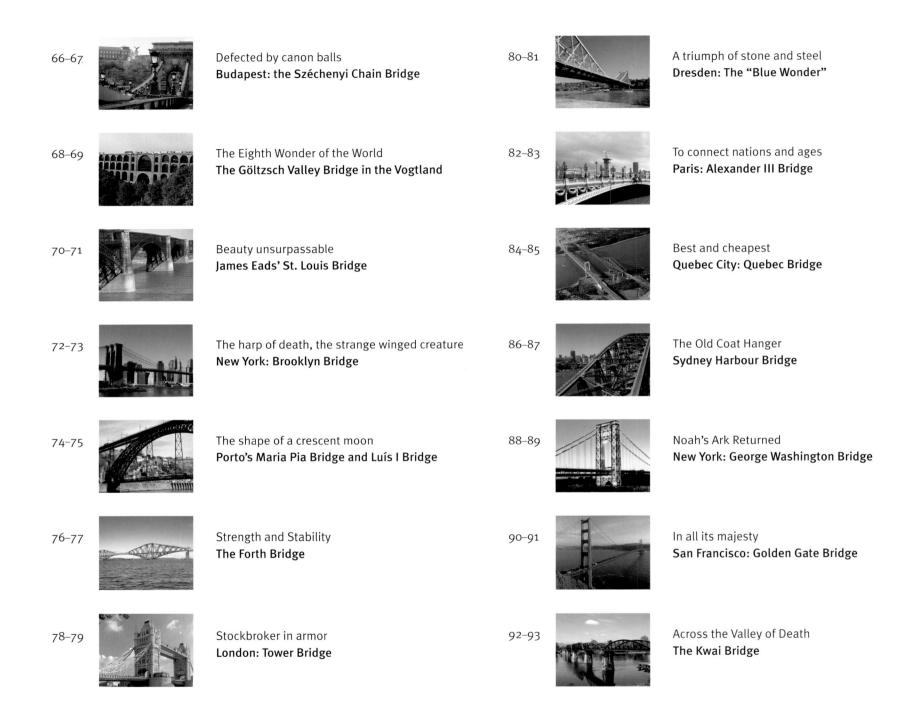

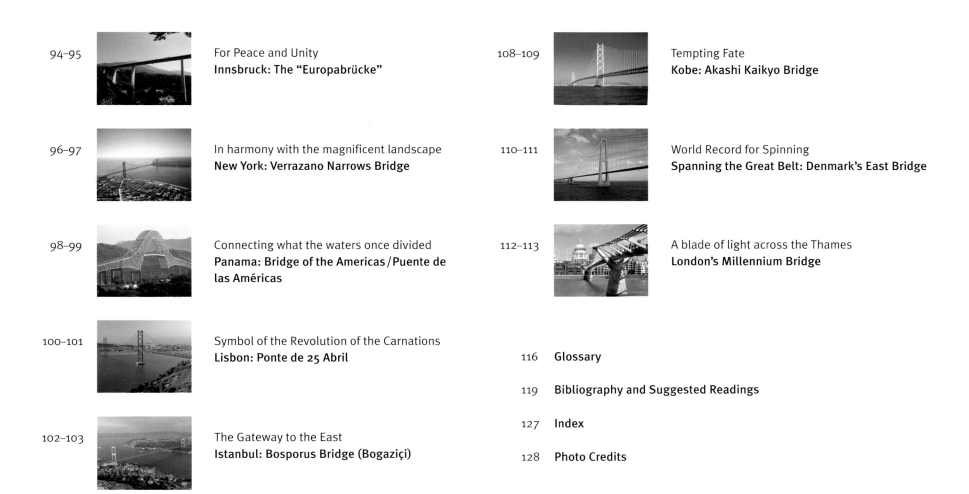

Daring constructions with a tendency to collapse: only when steel became widely available towards the end of the nineteenth century did railway bridges in America become safer.
William Henry Jackson, Dale Creek Bridge, Union Pacific Railway, *c.* 1870

Bridges have Meaning

"Bridge building is more efficient than any other form of architecture as it is accessible to everyone. Bridges are accessible to people who are not interested in art. One single gesture transforms nature and gives it order."

Architect Santiago Calatrava, 2001

There is no doubt that bridges have played a part in history. Sometimes, the role has been a connector of people, and other times a divider of people. Bridges have been used as defensive structures, such as the Kapell Bridge in Lucerne, keeping the enemies at bay. But bridges have also been used as living invitations to integration and interaction, like the Europabrücke in Innsbruck. Whatever the purpose behind the bridge may have been, the bridges themselves had meaning for the people who used them. As Santiago Calatrava suggests, even the most primitive people felt a connection with the inanimate object beneath their feet called a bridge, because it aided them in their lives in some way.

The concept of a "bridge" has many meanings today. Fortunately, the prevalent modern meaning focuses on the more positive role bridges have played in the past. Bridges now bring people across this blue planet of ours together, in every conceivable way. Figuratively, we think of diplomatic talks and negotiations between opposing factions, or courageous and ethical conduct by individuals wanting to bring about reconciliation between nations. We might also think of international sports and music events or the presentation of awards. Modern

telecommunications—direct dial telephone calls, faxes, e-mails, and the Internet—also help to forge links. The world was never as small or as readily contactable as it is today thanks to satellite technology.

The future also includes the kind of bridges Santiago Calatrava likes to build. We will see no letup in competitive bridge building: whose is the longest span, the most daring construction or the most unusual design? Japan's record-breaking Akashi Kaikyo Bridge will be superseded in every respect by the proposed 3.4-km-long bridge over the Strait of Messina (120 m deep) or one between Spain and Morocco that would be 14 km long. The Chinese-American engineer Tung-Yen Lin is interested in building intercontinental bridges and has proposed a link between Alaska and Russia, a distance of some 80 km. Professor Lin's basic need to overcome physical obstacles is as old as humanity itself. Our ancestors first expressed it in prehistoric times when a log was laid across a stream for the first time. Lin's proposal to build intercontinental bridges has an equally long tradition.

The Greek historian Herodotus (*c.* 485–425 B.C.) described how the Persian kings Darius I and Xerxes I connected Asia and Europe using pontoon bridges in

9

Karl Theodor von Dalberg, prince primate of the Rhenish Confederation, welcomes Emperor Napoleon I near the bridge across the river Main at Aschaffenburg. Oil painting by F.F. Bourgois and Jean Baptiste Debret, Musée National du Château de Versailles, 1812

The *Bridge at Langlois (Washerwomen)*, oil painting by Vincent van Gogh, Rijksmuseum Kröller-Müller, Otterloh, 1888

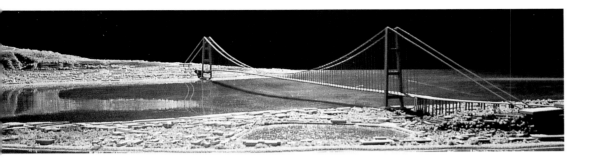

From 2011, the world's longest suspension bridge
is to connect Sicily and the Italian mainland. Model
of the bridge proposed for the Strait of Messina:
3.4-km-long, with towers 376 m high

513 and 546 B.C. respectively. Once the Bosporus and
the Dardanelles had been crossed, the Spartans and the
Athenians were to be destroyed on their native soil. In
55 B.C., Julius Caesar wrote in his *Gallic War* (IV, 16–17):
"As I saw how readily tempted the Teutons could be to
invade Gaul, my plan was best served by making them
fear for their own territory and by showing them that the
army of the Roman people was able and dared to cross
the Rhine…It was my belief that to do so by ships alone
would be insufficiently safe…Although it was apparent
that to construct a bridge would entail the greatest difficulties on account of the breadth, swiftness and depth
of the river, I accepted that I had to make the attempt."
In late antiquity, the Middle Ages and the modern era,
too, the construction, crossing or destruction of bridges
has crucially influenced military campaigns and political

situations. Think of the Emperor Constantine vanquish-
ing his rival Maxentius in 312 A.D. on the Milvian Bridge
near Rome, thus helping to promote Christianity; think
of Emperor Napoleon I and Karl Theodor von Dalberg,
prince primate of the Rhenish Confederation, meeting
on the bridge over the river Main at Aschaffenburg
(1806) before going on to defeat the forces of Prussia
and Russia; think of the battles that raged around the
bridges at Arnheim and Remagen during World War II;
think also of the bridge at Mostar during the Bosnian
civil war (1993). Neither is there a shortage of everyday
tales about meetings on bridges between pilgrims,
knights, merchants, riffraff, cut-throats and highwaymen
from the four corners of the earth. Sometimes the statuary placed on bridges was even regarded as a political
manifesto.

Among the *Bridges that Changed the World* are some
that became famous only after painters painted them,
poets eulogized them, singers sang about them or film
directors made a film about them. Who does not know
van Gogh's *Bridge at Langlois* (1888), or hasn't heard
of William McGonagall's *Tay Bridge Disaster* (1879), the
French song *Sur le pont d'Avignon* or seen footage of
the *Opening of the Williamsburg Bridge* (1904), or the
motion pictures *The Bridge on the River Kwai* (1957) or
Lovers on the Pont Neuf (1991)?

By examining more than fifty bridges from Europe
and Asia, America and Australia this book illustrates
everyday cultural and historical events without neglect-
ing the fate of those who commissioned them, the
careers of those who built them or their technological
achievements.

Bernhard Graf
May 2002

11

Built by the devil
In Exmoor National Park: Tarr Steps

"Tarr Steps, a wonderful crossing of Barle river, made (as everybody knows)
by Satan, for a wager."

R.D. Blackmore, *Lorna Doone*, 1869

According to locals, Old Nick himself built Tarr Steps, a "clapper" bridge northwest of Dulverton in the English county of Somerset. For that very reason no one was allowed to cross it—let alone bask in the sun on its hot granite slabs during the summer. This old tale high-lights the fears of local residents in centuries past when Exmoor was still covered by impenetrable forest. The Romans avoided it; in the Middle Ages, it became a royal hunting preserve; and as late as the seventeenth century, it was still the haunt of a family of freebooters immortalized in the Victorian novel *Lorna Doone*.

The 55-meter-long bridge must have seemed a real enigma. Who raised these stone blocks up to a meter above the surface of the river? Who put the slanted cutwaters in place? Who laid the seventeen blocks that form the walkway? These questions were all the more perplexing given that the massive slabs, some of them measuring up to 2.59 m long and 1.52 m wide, were not quarried in Exmoor. In the imagination of the locals, only the Devil himself could have been responsible for such an inconceivable feat.

Even today, visitors are amazed when they come ac-ross this construction, whose weight is the only thing holding it together. Probably derived from the medieval Latin and Anglo-Saxon word for "heap of stones," this type of bridge is known in English as a "clapper" bridge. Archaeologists are in the dark, however, when it comes to identifying the exact age of Tarr Steps or similar bridges, which is not surprising considering that they were built by pre-historic man in both the Bronze Age and the Iron Age, as well as during the early medieval period. Similar, albeit far more sophisticated, stone slab bridges were also found in ancient China. Wu Wung had one of them built across the Min river in the province of Szechwan around 1040 B.C. The bridge across the 400-meter-wide Ba river north of Xi'an stands to this day. According to Xu Jian, it was built during the Han dynasty (206 B.C.–219 A.D.). The Venetian merchant and traveller Marco Polo (1254–1324) wrote admiringly about the ancient stone slab bridges he saw in Chang-an: "They were of a length found nowhere else in the world." This was particularly true of the An Ping Bridge (1138–51) between Anhai and Shuitou during the reign of the Song emperor Gaozong (reigned 1127–62). At 2,500 meters long, it would have been unimaginable even for the builders of Tarr Steps, whether pre-historic or medieval.

At 2.5 km long, the Chinese An Ping Bridge between Anhai and Shuitou is the world's longest stone slab bridge, 1138–51

Like the Tarr Steps, the Postbridge near Princetown is a "clapper" bridge

The enigmatic Tarr Steps, an ancient "clapper" bridge across the river Barle in Somerset, England

In the Land of the Aryans and Khorasan
The beam bridges of Afghanistan

"The giant raised a hill onto his shoulders and cast it into the river. 'I want to see how big the hill is,' cried the prince, starting to run over the hill that formed a bridge across the river. He ran until dawn and then the whole of the next day before reaching the other side. Only in this way is it possible for a mortal being to cross the river into the realm of the fairies."

A river crossing as described in the
Afghan fairy-tale *The King with Forty-One Sons*

Ancient beam bridge located east of
Qala Panji in Afghanistan

Just as native North Americans were able to make use of the 32-meter-high Landscape Arch in present-day Utah to cross a mountainside, in the Afghan tale, the king's eldest boy uses a natural bridge to cross a river. In his case, however, the crossing was not created by erosion, but by a giant. It also happened to be the way to the land of the fairy queen, Shah Bolshah Pari, whom the prince hoped would be able to cure his ailing father. Figures in other Afghan tales were not quite so lucky: the king's three sons leading a caravan in the tale *The Prince with the Magic Horse*, for instance, are always on the lookout for fords where their camels and horses can cross rivers in safety.

Rivers played an important role in the history of Afghanistan's Stone Age inhabitants, the Aryan peoples of antiquity, as well as the Ghaznavid and Ghurid rulers in medieval Khorasan. They offered the people some protection from their enemies, but the Persians, Greeks, Huns and Mongols eventually conquered the fiercely disputed territory and ravaged it. Rivers also proved to be obstacles for the people they protected, cutting off trade and military traffic. To increase long-distance trade and ease of movement for their own military expeditions, the

region's peoples built beam bridges such as the one that has survived east of Qala Panji.

Tree trunks were laid across rivers and secured at either side using boulders, while branches and flat stones were used to make the walkway. A large span could not be created using such a technique, so to bridge a wide river, local builders had to take their inspiration from the advanced and densely wooded Chinese provinces to the northeast. As long ago as 305 B.C., the Chinese prince Chao Hsiang had a 2,000-foot-long wooden bridge built across the Wei river. Forty-eight years later, at the command of the ruler Chao Hsiang from Ch'hin, a similarly sized wooden pontoon bridge was built across the Yellow River. An ancient Chinese book called *Rivers* by Li Dao Yan (c. 472–527) even describes a wooden bridge near Paohan in the province of Gansu that was built during the Yixi era (405–418) on the cantilever principle and which had a span of 13 meters. These examples illustrate how Chinese bridge building techniques were far in advance of those found in the realms of the Aryans and Khorasans where war and destruction greatly hindered technological progress.

The Zamalong cantilever bridge near Xining is far superior to the wooden constructions found in ancient Khorasan

Trade route across an old bridge in the Wakhan valley in
Badakshan province, northeastern Afghanistan

From the Roof of the World
Nepal: Bridges across the Dudh Khosi and Trisuli Rivers

"The Prince struck the ground three times with his staff and pulled on his iron gloves. In an instant, the river divided and a path formed through its middle. On reaching the other side, the Prince said: 'Now flow as you did before!'"

Account of a wondrous river crossing in the Nepalese
fairy tale *Winning the Hand of Princess Bea Rani*

Footbridge of twigs and stones

How often must the Newari, the first people to inhabit the Himalayas, have been at a loss when confronted by a rocky ridge or a wide river? In numerous Nepalese fairy tales, travellers bemoan the long detours they have to make because of the lack of footbridges. How travellers wished they shared the magic powers of the king's son, whose ability to divide the waters of a seemingly impass-able river came to his aid as he made his way to win the hand of Princess Bea Rani. In another tale, a young lad searching for his mother comes across a river with no bridge, but his prayer is answered when the bamboo around him bends over the river, allowing him to cross.

Miracles could not always be expected and so the Newari had to grapple with the problem of building river crossings themselves. They buried bamboo poles in the ground until the lignin they contained had decomposed, leaving behind only tough cellulose fibres that could be made into durable and tension-proof ropes and hawsers. Three ropes, knotted together at close intervals, formed a primitive suspension bridge. A person could cross a ravine by placing one foot in front of the other on the middle rope, much like a tightrope walker, while holding on to the ropes on either side of him.

Using the most simple of suspen-sion bridges, animals and humans alike were borne across gorges in the Himalayas

As local bridge builders gained more experience, they were able to construct more stable suspension bridges such as that spanning the Trisuli river northwest of Kathmandu. A number of cellulose fibre ropes were tied tightly together and secured to the cliff face. Wooden planks were placed on top of them and also anchored to the cliff face. A form of railing was provided by an arrangement of ropes tied together on both sides of the walkway. This was the type of suspension bridge that the Chinese monk Fa Xian saw and described in 399 A.D. on his travels through the Licchavi kingdom (*c.* 300–750 A.D.). He noted that the wooden walkway resting upon the cellulose fibre ropes caused them to sag badly, thus making it difficult to walk across the middle of the bridge. This prompted local bridge builders to introduce the Chinese invention of the rope bridge. A roughly horizontal wooden walkway was attached to slack ropes over a gorge. Only short distances could be spanned in this way, of course, which was also true of the wooden cantilever bridges found on the "Roof of the World" such as the one in the Dudh Khosi valley in northeastern Nepal.

A suspension bridge across the River Trisuli in Nepal

16

Rather insecure in appearance: a wooden cantilever
bridge in Nepal's Dudh Khosi valley

At the Emperor's Command
The Aqueduct at Segovia   *c.* 100 A.D.

"Water is conducted in three ways: through conduits in masoned joints, through lead or earthenware pipes... Water from earthenware pipes is healthier than that conducted through lead pipes as the lead appears to be detrimental to health because it is a source of white lead... Where the supply reaches a city's walls, a reservoir is to be constructed."

Marcus Vitruvius, *De Architectura*, end of the first century B.C.

**Titus Flavius Domitian
(51–96 A.D.)**

c. **51 A.D.** born the youngest son of Emperor Titus Flavius Vespasianus (reigned 69–79 A.D.) and Flavia Domitilla
70 Roman praetor
81 following the death of his brother, Titus, whom he allegedly had poisoned, he was made Emperor on September 14
82 renovation of the Capitol in Rome
war against the Chatti
establishment of a limes
89 war against the Marcomanni governor A. Lappius Maximus suppresses the revolt by Saturninus
92 victory over the Sarmatians and Suebi
96 a praetorian prefect murders him on September 18 in a conspiracy involving his wife Domitia Latina

Aqueducts were nothing new. Even in the days of Emperor Augustus (reigned 30 B.C.–14 A.D.), architects were able to study the construction of aqueducts, or bridges that carried water across great distances, in *De Architectura*, by the Roman architect and engineer Marcus Vitruvius. Nearly seventy years after Augustus, Emperor Titus Flavius Domitianus (reigned 81–96 A.D.) commanded that the Roman town of Segovia be connected by a 17-kilometer-long aqueduct along a shallow gradient to Riofrío, the confluence of streams at the foot of the Guadarrama mountains.

The builders, who studied the manual written by Marcus Vitruvius, were as aware of the detrimental effect of lead pipes as of how to construct a reservoir, in which the water was collected and purified. A number of holding tanks allowed priority distribution of Segovia's water supply: first and foremost came the public baths and water wells; secondly, private baths; thirdly, private households. To ensure a gradual descent, a two-tiered aqueduct of Guadarrama granite blocks was built up to a height of 28.9 meters without the use of mortar and lifting cranes. The governor of Segovia had niches built into either side of the uppermost tier of arches, initially reserved for busts of the Emperor Domitian. Between the tiers of arches, an inscription of gilded bronze welcomed or bid farewell to travellers. The positions of the dowel holes reveal the wording on a supporting wall that measured 17.5 by 1.35 meters: "At the command of the Emperor Nerva Trajan, Pontifex Maximus, father of the fatherland, in his second year as tribune and consul, Publius Mummius Mummianus and Publius Fabius Taurus, principal magistrates of the citizens of Segovia, have restored this water supply." Trajan, emperor by adoption (reigned 98–117 A.D.), had the aqueduct renovated in the late autumn of the year 98 A.D. and had the old busts replaced by his own. Today, statues of the Madonna Nuestra Señora del Carmen (1520) welcome visitors to Segovia. The words of an English visitor in 1812 remain just as valid today: "If such an aqueduct were to be found in my country, it would be behind glass."

Grabado de G. Doré.

The Roman aqueduct at Segovia, as seen by Gustave Doré (1832–83)

Supported by 119 arches over a length of 728 meters, the aboveground section
of the aqueduct crosses outlying areas of Segovia; built *c.* 90–98 A.D.

To the Generations of the World
The Alcántara Bridge   c. 100 A.D.

"I, Caius Julius Lacer, famed for my divine art, leave this bridge for eternity and all the centuries of the world. Also a temple has been constructed at the will of the divine, Roman emperor in order to bring good fortune and blessings to this great bridge so that its burden may not ravage it."

Epitaph of Caius Julius Lacer on the temple
overlooking the Alcántara bridge, c. 100 A.D.

Marcus Ulpius Trajanus (53–117 A.D.)

53 A.D.   born September 18 in Italica (northwest of Seville) in the southern Iberian province of Baetica
97   adopted by Emperor Nerva (reigned 96–98 A.D.); named provincial governor of Upper Germany; established the provinces of Ulipa Traiana (Xanten) and Ulpia Noviomagnus (Nijwegen)
98   elevated to Emperor on January 28
105   conquered the Dacian Dekebal
107–112   construction of Trajan's Forum in Rome by Apollodorus of Damascus; erection of the 38-meter-high Trajan's Column
113   start of the Parthian War
114–116   the Roman Empire reaches its greatest extent
114   awarded the title of "optimus princeps" by the Senate
117   Jewish rebellion in Egypt and Cyrenaica; died on August 7 in Selinus, Asia Minor

Roman master bridge builder Caius Julius Lacer wanted it known in posterity that, around 100 A.D., the Emperor Trajan (reigned 98–117 A.D.) commanded him to build a granite bridge across a steep and narrow valley of the river Tagus (Tajo), a bridge to awe the "centuries of the world." Fifty-seven meters high and 194 meters long, the Alcántara even exceeded the dimensions of the Pont du Gard. With six semi-circular arches raised without the use of mortar to a height of 28.8 meters, it was among the largest bridges in the Roman empire—which itself had reached its greatest extent under Trajan. In addition, Lacer gave the bridge a 14-meter-high triumphal arch that bore the following inscription: "To the divine Emperor and Caesar Trajan, to Nerva's son, the greatest bridge builder, tribune for eight years, emperor and consul for five, the father of the fatherland."

Trajan had a monumental structure of such scale built here, near the Roman town of Norba Caesarea, to facilitate the transport of raw materials mined in the Roman provinces of Astures and Cantabri across the Tagus en route to the sea where they were shipped. The bridge was situated on the route of the military road leading from the northern garrison of Legio Gemina and from the towns of Asturica (Astorga) and Salamanca to the Roman province of Baetica, where Trajan was born in Italica in 53 A.D., and then on to the sea at Gades, today's port city of Cádiz.

Throughout history, the Alcántara bridge would fascinate and serve many masters. The Moors called Trajan's monumental structure "Al Cantara," meaning the Bridge, which would become the name of the village in the province of Cáceres (Extremadura), replacing the name Norba Caesarea. Emperor Charles V (reigned 1519–56), as king of Spain, had battlements added to the triumphal arch and his imperial eagle affixed to it in 1543 as an affirmation of his reign. The Spanish mystic and Franciscan friar, Saint Peter of Alcántara (1499–1562), reflected on the subject of "crossing the bridge," inspired by the great Roman bridge at his birthplace, Alcántara.

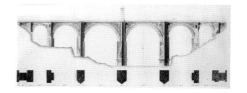

A cross-section view of the Alcántara Bridge, with
28.8-meter-high arches raised without mortar

The Alcántara Bridge, built under the Roman Emperor Trajan
(reigned 98–117 A.D)

Formerly the Pons Aelius

Rome's Ponte S. Angelo c. 134 A.D., 1667–71

"In Rome at the bridge that outclasses every other, that of Hadrian...whose remains I, too, view with reverence, there stood a roof that rested upon forty-two marble columns with horizontal entablature. The roofing was of bronze and afforded splendid decoration."

Leon Battista Alberti, *De re aedificatoria*, Lib. X, 1450/51

Publius Aelius Sergia Hadrianus (76–138 A.D.)

76 A.D. born January 24 in Italica
85 orphaned, adopted and raised by his relative Trajan
108 consul
117 succeeds Trajan as Roman Emperor on August 11, peace treaty with the Parthians
118–138 construction of the Villa Adriana at Tivoli
120–125 renovation of the Pantheon in Rome
122 reinforcement of the limes in Upper Germany
122–128 construction of Hadrian's Wall in the province of Britannia
132–135 war in Judea against Bar Kokhba
136 design of the Temple of Venus and Rome adopted by Lucius Ceionius Commodus
138 died July 10 at Baiae

Around the middle of the fifteenth century, the master builder, sculptor and painter Leon Battista Alberti (c. 1404–72) imagined the Aelius Bridge of Roman antiquity to be a covered structure. The architects Antonio Averlino Filarete (c. 1400–69) and Andrea Palladio (1508–80) shared his view. Roman medals, however, show that all three were mistaken: the Aelius Bridge had no roof. Above its three main arches stood four pedestals supporting bronze figures of Victory, each holding aloft a wreath with which to glorify the Roman Emperor, Publius Aelius Hadrian (reigned 117–138 A.D.). Emperor and architect, he had the triumphal bridge built for himself. It served as a crossing from the Field of Mars, south of the Tiber, to his mausoleum (135–139 A.D.), today's Castel Sant'Angelo. Alberti was thus right to stress the prominence of the Aelius Bridge among the eight bridges spanning the river Tiber in Rome.

The Aelius Bridge acquired different functions over time. After 1527, Pope Clemens VII (reigned 1523–34) had statues erected of the apostles and of Rome's patron saints, Saint Peter by Lorenzo Lotti (1490–1541) and Saint Paul by Paolo Taccone (died 1477). In 1536, the sculptor Raffaelle da Montelupo (1505–66), in a

commission from Pope Paul III (reigned 1534–49), produced stucco figures of the patriarchs and four Evangelists for the ceremonial entry of Emperor Charles V (reigned 1519–56) into the city. They took the place of the statues of Victory that had long since disappeared. After 1671, the Aelius Bridge received its final set of decorative figures and additional bays by Giovanni Lorenzo Bernini (1598–1680) and his assistants. The bridge was graced by ten statues of angels bearing the instruments of Christ's suffering, signifying the meaning of Christ's passion to those approaching the Vatican City. The ancient Roman bridge had become the bridge of the Saviour and the Angels.

In 1717, the cleric and artist's son Domenico Bernini wrote: "The papal primacy is most reliably secured by the Pope exercising the office of a bridge builder, of a mediator between God and men—like a bridge connecting one bank to another."

One of the ten angels on the bridge

top left: Giovanni Piranesi's veduta from the mid-eighteenth century shows the elongated Ponte Sant' Angelo with St. Peter's in the background

Rome's Ponte S. Angelo built under the Emperor Hadrian around 134 A.D., enlarged by Bernini in the seventeenth century

An arch in flight
Zhao Xian, Hebei: The An Ji Bridge   605–17

"Its arch is gentle; its facing stones fit together unsurpassably. What lightness characterizes this arch in flight and its openings. A work such as this will endure for centuries."

Inscription on the An Ji Bridge

The An Ji, also known as the Zhaozhou Bridge, at Zhao Xian in the Chinese province of Hebei has endured longer than centuries. Apart from some repair work during the Ming (1368–1644) and Qing dynasties (1644–1911/12), this bridge has spanned the river Xiache unchanged for 1,400-odd years. For this very reason, it was named An Ji, or "Safe Crossing."

In Hebei province, it was said that the bridge had been built not by human hands, but rather heavenly beings at the command of the Jade Emperor. Decorative nineteenth-century reliefs recalled such talk and revealed just how fascinated the Chinese were with the perfection of their oldest stone bridge. The master architect and stonemason Li Chun built it during the reign of the Sui Emperor Yang-ti (reigned 604–17), who ruled China from his capitals at Chiang-tu and Lo-yang. Li Chun constructed it as a segmental arch with open spandrels, a remarkable achievement considering that it predates comparable bridges in Europe by some 700 years (e.g. Florence's Ponte Vecchio). The curvature of the arch of the An Ji Bridge is derived from a section of a circle measuring 55.4 meters in diameter. This allowed Li Chun to span a distance of 37.02 meters. Using X-shaped iron dovetails in the vertical joints of the arch stones, he was able to ensure the shear strength of the main arch, comprised of twenty-eight courses of masonry.

The bridge's unique durability is celebrated both in a famous folk-song and, since 1790, the fifty-fifth year of the reign of the Manchu Emperor Ch'ien-lung (reigned 1735–96), in an aria from the Peking opera depicting a collection of historical events and social wrongs, tales of heroic deeds and love, but also strange encounters. It was said, for instance, that the mythical beings of Zhangguolao, astride a donkey, and Chaiwangye, pulling a cart, could cross the An Ji Bridge together without causing it to collapse. This is all the more impressive given that as "luggage," each of them carries China's five highest mountains with him! The American Society of Civil Engineers also appreciated the strength and beauty of the world's oldest segmental arch bridge. In 1991, they voted the An Ji Bridge the twelfth international milestone of historic architectural importance.

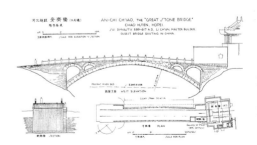

Design drawing for the An Ji Bridge

The world's oldest segmental arch bridge, the An Ji Bridge,
in the Chinese province of Hebei

Whence there is only one route
Puente la Reina: the Pilgrims' Bridge

"There are four ways to Santiago that converge into one at Puente la Reina in Spain... The routes by way of Ste-Foy in Conques, St-Léonard in Limousin and St-Martin in Tour join at Ostabat; once they cross the Cisa Pass, at Puente la Reina they meet the road that crosses the Somport Pass, whence there is only one route to Santiago de Compostela."

Codex Calixtinus, Book V, c. 1160

The Gothic "Santiago Beltza," the black St. James in the Church of St. James in Puente la Reina, reminds us that to this day Santiago de Compostela is the most significant place of pilgrimage alongside Rome and Jerusalem

The name "Puente de los Peregrinos," the "Bridge of Pilgrims," was not only heard in the medieval towns and villages of Navarre. Throughout Europe, and as far afield as the African kingdom of Nubia, pilgrims related how they had crossed that most famous of bridges on their pilgrimage to the shrine of St. James at Santiago de Compostela. According to the oldest accounts, pilgrims of all stations met at the bridge: princes and knights, bishops and monks, burghers and peasants who had travelled along the main French routes called the "Via Tolosana" (Toulouse), "Via Podensis" (Le Puy), "Via Lemovicensis" (Limoges) and "Via Turonensis" (Tour). In each other's company, they continued along the "Camino Francés" by way of Burgos and Léon to Galicia.

The cult of St. James began in the ninth century, prompted by the desperate struggle of the Christians against the Moors. Early pilgrims had to find other ways to reach the shrine of St. James, for the Puente de los Peregrinos was not built until the 11th century. A Romanesque structure, the bridge was built on the orders of Queen Doña Elvira Mayor of Castile, the wife of King Sanchos III Garcés of Navarre and Aragón (reigned

1000–35), or at the behest of her daughter-in-law Doña Estefanía of Barcelona, the wife of her son, García III Sanchez de Nájera (reigned 1035–54). The bridge lent its name to the spot, Puente la Reina, the Queen's Bridge. Six round arches resting on five piers with cutwaters and just as many overflows made a permanent river crossing possible. Knights of the Order of St. John of Jerusalem and Knights Templar secured it against robbers and highwaymen, the rabble and lapsing penitents.

The area around Santiago

Where there was no bridge or where one had been destroyed, pilgrims to the shrine of St. James had to use a staff to jump across a stream; to cross a wide river, however, a big detour was necessary, Heinrich Wölffli, Johann Haller, Berne

Countless pilgrims from all over Europe used the "Puente de los Peregrinos," the Bridge of
Pilgrims, to cross the river Arga en route to Santiago

In the Care of the "Bridge Masters"

Regensburg's Stone Bridge   1135–47

"When the year was 1135,/a great bridge of hewn stone was built over eleven years,/with fifteen arches and a bay,/a tower was raised upon its centre,/through which the Danube sped,/on its way to Austria and Hungary."

Hans Sachs, *Städtelob*, 1565

King Philip of Swabia (reigned 1198–1208), whose "Philip-pinum" of 1207 granted the townspeople of Regensburg extensive rights to the bridge; formerly situated on the north facade of the Central Tower, beginning of the thirteenth century. Museum der Stadt Regensburg

When Nuremberg's mastersinger Hans Sachs (1494–1576) honored Regensburg's Stone Bridge with his panegyric, it was still wholly intact. The cofferdams had already formed the foundation of the 336-meter bridge across the Danube for more than 400 years. Formed by two rows of piles driven into the riverbed, made water-tight with wattle, stones and clay, drained, and sealed with rubble, the cofferdams prevented erosion of the gravel bottom and protected the fifteen bridge piers. Time and again, the bridge had to be repaired and extended; a job that fell to the "bridge masters," who added the Black Tower in the twelfth century and reinforced it with a bridgehead comprised of ditches, turrets, and a drawbridge in 1383.

On the north facade of the four-storey, tin-sheathed Central Tower, built around 1200, likenesses of Queen Irene and her spouse Philip of Swabia (reigned 1198–1208) watched over the merchants crossing the bridge. A mask embedded in the wall was intended to protect the bridge and the town from evil spirits. Moreover, a winged lion provided a connection between Regensburg and Rome and Venice. City officials examined goods in transit in the Central Tower's wooden guardroom and

allowed merchants to pass only after payment of duties. Around 1300, the third tower, called the "Brücktor," was built. The townspeople of Regensburg used it as a "debtors' jail, which earned it the title of 'Debtors' Tower.'"

In Hans Sachs' day, merchants from Austria, Hungary, Russia, Venice, France and Brabant were greeted by a black imperial eagle on a yellow field affixed to the north side of the Black Tower, while on its south side there hung an image of King Oswald (St. Oswald). Those entering the tower knew they were entering the Imperial Town of Regensburg, a place with distant trading links. Visitors could also hear the sound of the heavy waterwheels from a number of mills: the fulling mill, the sawmill, the polishing mill, and the paper and spice mill that the "bridge masters" built in the fourteenth century between the bridge piers to take advantage of the power of the Danube.

Damaged by war in 1633, 1809 and 1945 and by ice in 1784, the "Black Tower" and "Central Tower" had to be demolished and massive repairs to the arches were made.

View of Regensburg's Stone Bridge in *Topographia Germaniae*, engraving by Matthäus Merian the Elder (1593–1650), 1644

Model of the Stone Bridge

Nowadays only fragments of Regensburg's Stone Bridge (1135–46) survive

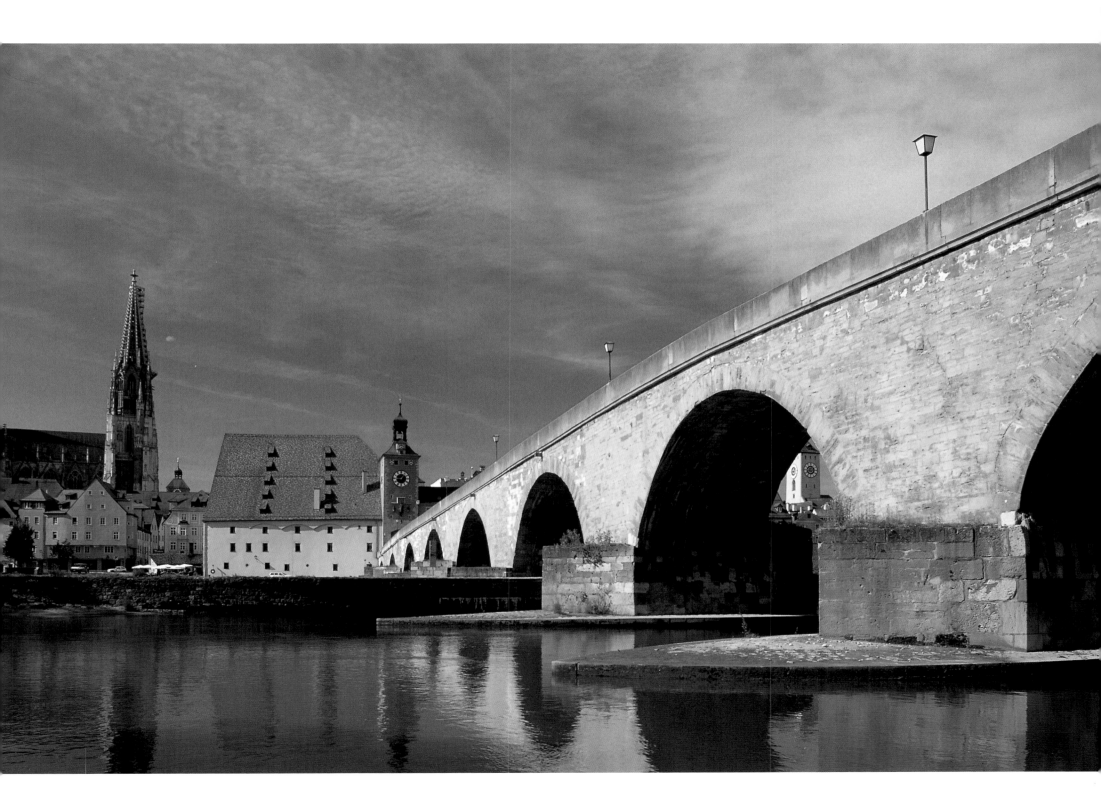

Called by God

Avignon: St. Bénézet's Bridge 1177–87

"The shepherd boy Bénézet, a slip of a lad with no possessions, has declared that he is able to build a bridge at the spot where neither God, Saints Peter and Paul nor Charlemagne nor any other has been able to do so. It would be a miracle ... I will believe him capable of building a stone bridge only when he moves a limestone boulder in my palace."

Legendary words of the Bishop of Avignon to the faithful, 1177

St. Bénézet (1165–1184)

1165 born in Hermillon near St-Jean-de-Maurienne in Savoy
1177 according to legend, God directed him to go to Avignon in the guise of a shepherd to construct a bridge across the Rhône
1184 died on April 14 in Avignon
1189 Pope Clemens III (reigned 1187–91) sanctioned the lay brotherhood of bridge builders founded by Bénézet, the "fratres pontifices"
1212 beatification
1233 canonisation; St. Bénézet becomes patron saint of Avignon, bridge builders and shepherds

An ancient Provençal legend relates that the shepherd boy Bénézet was indeed able to accomplish the task set by the Bishop of Avignon. In a display of amazing strength Bénézet dragged a huge piece of rock from the Bishop's palace to the spot on the riverbank where he planned to build the bridge over the river Rhône. The widespread tale of his supernatural abilities served its purpose; every vassal in Avignon now believed that divine assistance was involved and made charitable donations to secure the bridge's construction. The belief in divine assistance was so strong that the people of Avignon did not care whether Bénézet or a monk called Bénoit really built the bridge, according to the architect and restorer Eugène-Emanuel Viollet-le-Duc (1814–79).

The bridge at Avignon eclipsed the masterpieces of the bridge builders of antiquity. Not only was the stone bridge with twenty-two arches, held together without the use of mortar, built over the remains of a wooden Roman predecessor, but also it was the longest bridge in the medieval world, with an overall length of 920 metres. It was considered more elegant than its contemporary, the London Bridge, and more important. Until completion of the Pont St-Esprit in 1309, St. Bénézet's Bridge

provided the only Rhône crossing between the sea and Lyon. Bénézet did not live to see the completion of his divinely-inspired wonder of technology—as his contemporaries regarded it. He was buried in the newly consecrated chapel of St-Nicolas, situated on the bridge's second pier. After 1233, Bénézet's shrine became a place of pilgrimage. When Pope Clemens V (reigned 1305–14) moved the papal residence to Avignon on June 5, 1305, St-Bénézet's Bridge became a meeting place for negotiators and diplomats, senior clergy and princes. A medieval proverb claimed that on the bridge at Avignon, you would always find two monks, two donkeys and two harlots. St. Bénézet's Bridge remained world famous, despite Avignon's fall into obscurity following the departure of the Popes from the city in 1378, the loss of its saint to grave robbers and having all but three of its arches swept away by a high tide in 1665, thanks to the French folk song *Sur le pont d'Avignon*.

St. Bénézet shouldering a boulder as a reminder of his supernatural strength, Archives départementales, Avignon

The remaining arches of the "divine" St. Bénézet's Bridge at Avignon

Bridge of the Honorable Passage at Arms
Puente de Órbigo   thirteenth century

"Suero de Quiñones, the second son of the royal judge of Asturia...held a tournament beside the Bridge at Órbigo together with twelve knights and noblemen. Every knight or nobleman who passed by the bridge was obliged to tilt in plate armour and fight in single combat using iron weapons sharpened with a diamond until one of the combatants had broken three lances."

The Story of King John II, chapter 240, 1434

The chronicler at the court of King John II of Castile (reigned 1406–54) refers to the tournament staged by Don Suero de Quiñones in an attempt to win the noblewoman of his heart. In a written statement, he announced that he was in the throes of an inescapable passion. To show his love for the noblewoman, he wore a golden armband bearing the slogan, "If it does not please you to practice moderation, I tell you that you will be without happiness." On July 10, 1434, with permission from King John II, he declared a tournament in the name of St. James the Greater. To impress his noblewoman, he vowed that the jousters of this tournament would break three hundred lances. Only knights willing to enter the tournament were granted passage across the 110-meter-long Óbrigo Bridge.

For thirty days, the royal scribe, Pedro Rodríguez de Lena, took his seat in the stand that had been erected beside the Romanesque bridge, ready to observe and record the events of the jousting. Based on de Lena's records, the Franciscan friar Juan de Pineda reports that Suero de Quiñones equipped his rivals with the same weapons and paid their expenses. Sixty-eight knights from a number of European kingdoms announced their readiness to join single combat. The tournament at the Órbigo Bridge did not pass off harmlessly. A lance punctured the visor of a knight of the Holy Roman Empire and injured one of his eyes. Esberte de Claramonte of Aragon lost his life on the tournament field.

On August 9, Suero de Quiñones brought the tournament to a successful close. To the sound of a fanfare, he untied his finely embroidered armband and, together with his fellow jousters, galloped off to Santiago de Compostela. In gratitude to Jakobus he donated the armband that had belonged to the lady of his heart.

The tournament and the Bridge of the Honorable Passage at Arms were still a topic of conversation some 200 years later. In his parody on the romance of chivalry, Don Quijote de la Mancha, the Spanish playwright, novelist and poet Miguel de Cervantes Saavedra (1547–1616) lent the event and the Órbigo Bridge lasting fame when he had Don Quijote say to a jousting opponent: "Let no man tell me that the jousting of Suero de Quiñones, the hero of the tournament at the Órbigo Bridge, was a mere prank."

Don Quijote de la Mancha refers to the tournament at Órbigo Bridge in his attempt to convince a canon of the high ethical value of a tourney, Grandville, illustration for *Don Quichotte de la Mancha*, 1848

The Spanish playwright, novelist and poet Miguel de Cervantes Saavedra (1547–1616); his work immortalised Suero de Quiñones' tourney beside the Órbigo Bridge

Masons' marks from the thirteenth century indicate the age of the 110-meter-long bridge across the river Órbigo; the names of Suero de Quiñones and his fellow jousters are still visible.

Of Swiss Heroic Deeds

The Kapell Bridge in Lucerne 1300–33

"Like the Court Bridge, the Kapell Bridge shall also be decorated with painted panels; their content, however, shall be secular in nature, not sacred. They shall tell the story of our town and fatherland."

Resolution of the Lucerne Town Council, 1599

Hans Heinrich Wägmann (Wegmann) (1557–1628)

1557 born October 12 in Zurich
1584 paints choir panels for Eigental church
1589 receives full rights as a citizen of Lucerne, paints the town hall tower
1590 designs altarpiece for Lucerne's Court Church
1590–94 supplies Lucerne's Jesuit church with paintings
1611–16 completes panel paintings for the Kapell Bridge
1617 paints altar panels for the church in Greppen

Johann Caspar Meglinger (1595–c. 1670)

1595 born the son of a stonemason on August 15 in Lucerne
1626–35 paints "danse macabre" for Lucerne's Spreuer Bridge
1639 completes scenes from the life of St. Francis; commemorative portrait of the builder of Lucerne's Court Church; completes altarpiece for the Capuchin Church in Schüpfheim

The Lucerne Town Council waited seven years before it commissioned the local apothecary and town clerk, Renward Cysat (1545–1614), to create a series of images for the Kapell Bridge. They were then realized in 1616 by Hans Heinrich Wegmann (1557–1628), a miniaturist, glass and panel painter based in Zurich. He produced 158 Mannerist panel paintings inspired by the history of the Swiss Confederation and the town of Lucerne, the legends surrounding its patron saints, St. Leodegar and St. Maurice, and the patrons of the Court Church. In addition, Cysat and a council member, Hans Rudolf Sonnenberg, composed some explanatory verses. The decoration was designed to communicate the confidence and pride felt by the citizens of Lucerne and the Swiss Confederation.

The Kapell Bridge's original function was not to provide a venue for the illustration of heroic Swiss deeds. Built between 1300 and 1333, during the rule of the Hapsburgs, in the form of a 285-meter covered trestle bridge, utilizing both square stone piers and wooden piles, the bridge completed the line of defense afforded by the city walls that ran south and north of the river Reuss. An octagonal "water tower" further bolstered

the heavily fortified bridge. These measures were intended to repel the original cantons if they attacked from Lake Lucerne.

The Kapell Bridge lost its defensive function when Lucerne joined the Swiss Confederation in 1332 and after the Hapsburg Duke Leopold III (reigned 1365–86) had fallen at the Battle of Sempach (1386). Lucerne was now able to develop freely. Between 1400 and 1408, the wooden Spreuer Bridge, with its oriel chapel, was built just north of the Romanesque river crossing. After 1626, Johann Caspar Meglinger (1595–c. 1670), a local portraitist and painter of historical scenes, designed a "danse macabre" of 65 panel paintings to decorate the Spreuer Bridge. In times of battle, they were to remind men of the transience of earthly delights and to console ordinary folk with the thought of glory in the life hereafter. In August of 1993, fire largely destroyed the Kapell Bridge, Europe's oldest wooden trestle bridge, leaving the younger Spreuer Bridge unharmed.

Paintings in the interior of the Kapell Bridge

A panel from the "Danse macabre" cycle on the Spreuer Bridge by Johann Caspar Meglinger, 1626–35

Lucerne's Kapell Bridge, 1300–33, shortened to 222 meters in length during the nineteenth century, restored following the 1993 fire

A Pact with the Devil
Cahors: The Valentré Bridge 1308–55/78

"The engineer had grown weary of the slow progress. To gain the help of the Devil, he entered a pact with him. Shortly before the bridge was completed, he hit on an idea to save his soul: he gave the Devil a riddle and asked him to fetch some water for the last lot of mortar. When the Devil found his task was impossible and realised he had lost the wager, he took his revenge each night by breaking off a piece of the central tower that the masons had to replace."

Legend surrounding the Valentré Bridge, fourteenth century

Pope John XXII
(c. 1244/5–1334)

c. **1244/5**   born Jacques Duèse, the son of a wealthy bourgeois family in Cahors
educated by the Dominicans
c. **1300**   named bishop of Fréjus
1309–10   named chancellor to Robert of Anjou, King of Naples (reigned 1309–43)
1310   named bishop of Avignon
1312   named cardinal bishop of Porto
1316   elected Pope on September 5 in Lyons
founded a Carthusian monastery and a university at Cahors
1324   excommunicated king Louis IV of Bavaria on March 23 (reigned 1314–47)
1327   derided as "the priest Jacob of Cahors" by Holy Roman Emperor Louis IV
1334   died at Avignon on December 4

When the architect Paul Gout (1852–1923) renovated the Valentré Bridge in 1879, the residents of Cahors told him the legend surrounding its builder and his pact with the devil. It explained the missing section of the bridge's central tower. To keep the old tale alive, Gout asked the sculptor Calmon to chisel a figure of the Devil trying to remove a stone from the tower. Gout then placed the figure in the right spot, giving rise to the name "Devil's Bridge" or "Satan's Tower."

The bridge that caused the medieval burghers of Cahors to fear for their mortal souls was inspired by an earlier, more earthly, fear. Their wealth, accumulated in business dealings with Lombardy moneylenders and traders and merchants from near and far, was continually under threat. No one was more aware of this than the local banker's son Jacques Duèse, the bishop of Fréjus. On June 17, 1308, he persuaded the town's chief magistrates to build a fortified bridge across the river Lot, but he was unable to follow the construction work as his office of chancellor at the court of king Robert the Wise (reigned 1309–43) took him to Naples. He would later attain the papacy as John XXII (reigned 1316–34) but he did not live to see the Valentré Bridge completed.

The 138-meter-long bridge with its six 16.5-meter-high arches was not completed until 1355.

The bridge was well designed to defend the town in times of war. Its sharp cutwaters faced upriver and had embattled parapets from which the bridge's defenders were to be able to take the enemy under flanking fire. This was thought to be especially important following the French defeat at Crézy-en-Ponthieu (1346). Facing threat from the English and their Black Prince (1330–76), there was an urgent need to erect an outer bailey and towers with embattled parapets and machicolations. The bridge, with its portcullised and machicolated towers and double curtain walls, withstood many of the English attacks—but to no avail. In 1360, Cahors was ceded to Edward III (reigned 1327–77) by the treaties of Brétigny and Calais. This once wealthy trading center found itself on the brink of economic ruin when the occupying English forces withdrew in 1450.

Calmon's devil atop the middle tower of the Valentré Bridge

After the Battle of Crézy (1346), the English posed a threat to Cahors and the Pont Valentré, miniature in an account by Froisart, Bibliothèque Municipiale de Toulouse

The fortified Valentré bridge with its 40-meter-high towers defended
the river Lot and the town of Cahors, built 1308–55/78

800 florins' gulden rent a year

Florence: Ponte Vecchio   1345, 1564/65

"Taddeo Gaddi was commissioned to produce a model and drawings for the Ponte Vecchio. He was to make this bridge as solid and lovely as possible. Accordingly, he shunned neither cost nor effort and built it with mighty piers and splendid arches...on account of which it was sufficiently strong to carry twenty-two shops on either side. This is of great benefit to the town in that it receives 800 florins' gulden rent each year."

Giorgio Vasari, *The Lives of the Most Eminent Italian Architects, Painters and Sculptors*, 1550

Taddeo Gaddi
(c. 1300–1366)

c. **1300**   born
godchild of Giotto di Bondone
(1266–1337)
1341–66   paints frescoes in Florence: S. Miniato al Monte; Bellacci; S. Andrea in S. Croce; S. Spirito, Annunziata; S. Maria Novella
1366   dies in Florence

Giorgio Vasari
(1511–1574)

1511   born July 30 in Arezzo
1555–74   paints frescoes in Florence's Palazzo Vecchio
1560–74   construction of the Uffizi in Florence
1562   construction of the House of the Order of Santo Stefano in Pisa and of the Badia in Arezzo
1571–73   paints the Sala Regia in the Vatican
1574   dies June 27 in Florence

No less a figure than the father of art history, the renowned painter, architect and writer Giorgio Vasari (1511–74), attributed the Ponte Vecchio, Florence's Old Bridge, to Taddeo Gaddi (*c.* 1300–66), a pupil of Giotto and an artist known mainly for his frescoes and panel paintings. In 1333, the Arno swept away the predecessor of the Ponte Vecchio except for two central piers. It was another twelve years before the city fathers were able to start re-building the bridge because the height of the riverbank had to be increased to prevent more flooding.

Aesthetics, stability and commerce were the guiding criteria for Gaddi's design. The bridge's beauty and durability, praised by Vasari at the time of the 1557 flood, were the result of three extremely shallow segmental arches (3.9 to 4.4 meters in height) that span the 100-meter-wide river. This type of bridge was unknown in Europe at the time; however, it is not impossible that the architect knew of the An Ji Bridge in the Chinese province of Zhao Xian (seventh century). Florentine contemporaries such as Francesco Balducci Pegolotti, a representative for Bardi, maintained close trade relations with the Chinese rulers of the Yüan dynasty (1280–1368). The new Ponte Vecchio was to include

enough space for forty-four shops. Those on the bridge's Romanesque predecessor (1177/78) had produced a good return. This source of income dried up in 1495 when the city fathers sold the shops to private merchants. In 1564/65, Grand Duke Cosimo I Tuscany (reigned 1537–74) dramatically changed the appearance and function of the Ponte Vecchio. He instructed Giorgio Vasari to add a corridor to the east side of the bridge and to construct a covered walkway above it, giving Cosimo I direct access from his residence, the Palazzo Pitti, to the Uffizi and the Palazzo Vecchio, the administrative and political heart of Florence. In 1593, on the orders of Ferdinando I de Medici (reigned 1587–1609), only wealthy gold and silversmiths were allowed to set up businesses on the Ponte Vecchio. He had found the stench from the previous businesses unbearable when crossing the Arno by means of the enclosed walkway.

top left: Bernard Bellotto's *View of the Ponte Vecchio over the Arno*, 1742, Beit Collection, Russborough

Doing business on the Ponte Vecchio, anonymous artist, Grassi Collection, 1850

According to Giorgio Vasari, the Ponte Vecchio originally had 44 shops

Built by a "raging dog"
Verona: the Scaliger Bridge 1354–56

"The Castelvecchio Bridge is significant on account of the length and shallowness of its arch that possibly has no equal. This bridge is no longer used by traffic, being open instead once a year only for a procession."

Friedrich Wilhelm von Erdmannsdorff on Verona's Scaliger Bridge, 1765

Cangrande II della Scala (1332–1359)

1332 born June 8, the son of Mastino II della Scala (reigned 1329–51)
1350 married Elisabeth (1329–1402), daughter of the Wittelsbach emperor on November 22 in Verona
1354 had two castles refurbished at Montecchio Maggiore
1354–56 construction of the military bridge across the Adige
1354–59 construction of Verona's Castelvecchio
1359 murdered by his brother Cansignorio (reigned 1352–75) on December 14

Cangrande della Scala

In the company of his prince, Leopold III of Anhalt-Dessau (reigned 1751–1817), Baron Friedrich Wilhelm von Erdmannsdorff (1736–1800) visited Verona in November 1765. The city on the river Adige then belonged to the Republic of Venice under the leadership of doge Alvise Mocenigo IV (reigned 1763–78). By then the Scaliger Bridge had long ceased to fulfil its original function, to form an extension to the defenses of the Castelvecchio, Verona's castle, beyond the line of the river Adige. It had been the resolve of the powerful house of Scaliger to secure the Adige and defend Verona since Alberto I della Scala was elevated to the position of chief magistrate of the city (reigned 1277–1301). In 1311, his sons Alboino and Cangrande I were even made imperial vicars by the Holy Roman emperor Henry VII (reigned 1308–13) and extended their rule beyond Feltre, Belluno and Treviso. Following the death both of his father and uncle Alberto II, Cangrande II assumed control in 1352. Two years before in Verona, he had married Princess Elisabeth, the daughter of Holy Roman emperor Louis IV (reigned 1328–47). Poets far beyond the confines of northern Italy sang the praises of Cangrande II. More than anyone else, he was concerned with the

defense of his royal seat and, after 1354, he had Verona's castle, the "Castello di San Martino in Acquaro," and the defensive bridge across the Adige built.

Cangrande II hoped that the castle and bridge would secure the town against the ambitious Visconti, against whom he had formed an alliance with Charles IV (Holy Roman emperor 1355–78). Cangrande II always felt threatened—even by members of his own family, hence his brutal suppression of the uprising led by his illegitimate half-brother Fregnano. His contemporaries thereafter referred to him as the "raging dog."

In the end, however, the strong defenses of Verona's castle and the Scaliger Bridge were of no avail to him. In 1359, his own brother Cansignorio murdered him. The remaining members of the Scaliger family did not fare much better. Cansignorio had his brother Paolo murdered so as to secure the succession of his illegitimate sons Bartolomeo and Antonio. The one became the victim of a treacherous murder, while in 1387 the other lost his royal seat with the Scaliger Bridge to Gian Galeazzo Visconti (reigned 1378/85–1402) in a war against the Signori da Carrara. His son, Gian Maria, was finally obliged to cede Verona to Venice.

Bernardo Bellotto, *View of Verona* showing the Castelvecchio and Scaliger Bridge (detail), 1745/46, Philadelphia Museum of Art

The Scaliger Bridge's lengthening spans are apparent here: 24 m, 27 m and 48.7 m. On April 25, 1945, the retreating "wehrmacht" destroyed the bridge. Reconstruction work was completed in 1958

Drowned in the waters of the Moldau

Prague: Charles Bridge   1357–99, 1464, 1591

"King Wenceslaus held a large firebrand in his hand and trailed it across the loins of Johannes... until he lost consciousness... The king then ordered that the honorable doctor, tormented to death, should be bound hand and foot, and, with a wooden gag in his mouth, taken by night to the river Moldau into which he was pitched from the bridge... Johannes drowned in its waters and thus his days were ended wretchedly."

Johannes Nepomuk, sculpture on the Charles Bridge, designed by Matthias Rauchmiller, by Johann Brokoff and cast by Wolfgang Hieronymus Heroldt, 1683

Johannes von Jenstein, archbishop of Prague,
in his account to Pope Boniface IX, 1393

**Peter Parler
(1330–1399)**

1330   born in Schwäbisch Gmünd
1353   apprentices to his father in
Cologne
1353   calles to Prague as master
builder of the cathedral
1353–56   completes Prague's
cathedral choir
1357   constructs Charles Bridge
across the Moldau
builds choir of All Saints Chapel
Royal on the Hradschin
1360   constructs choir of the
church at Kolín on the Elbe
1377   builds tomb of king Otakar I
(reigned 1198–1230)
1385   constructs choir stalls in
Prague's St. Vitus' Cathedral
1392   builds nave of Prague's
St. Vitus' Cathedral
1399   dies on July 13

There was great indignation in Perugia in July 1393 when Pope Boniface IX (reigned 1389–1404) learned of vicar-general Johannes von Pumuk's murder. Word soon spread among the subjects that the vicar-general and confessor of Queen Sophie of Bohemia (1376–1425) had been pitched from the Charles Bridge because he had not divulged to the jealous King Wenceslaus IV of Bohemia (reigned 1363–1419) whether the Queen had confessed to being unfaithful. The ensuing power struggle ended with the resignation of the archbishop of Prague, Johannes von Jenstein (in office 1379–96) and the dethronement of the King in 1400. Prague also lost the happiness associated with the history of the Charles Bridge.

Charles of Luxembourg, who ruled Bohemia as Charles IV (1346–78), commissioned Peter Parler (1330–99), the master builder of Prague's cathedral, to construct the Charles Bridge around 1357. Parler began to demolish the Romanesque flood-damaged Judith Bridge, leaving intact the "Small Side" Tower of 1172; as a complement to it, he built the "Old City" Tower. When the royal confessor was pitched into the Moldau, the Charles Bridge was still unfinished. The higher of the towers, the

"High Tower," was added in 1464; not until 1591 did the "Small Side" Tower take on its present form. Numerous figures were carved throughout the years to decorate the bridge, including: St. Vitus, the patron saint both of Prague Cathedral and of the line of Luxembourg in Bohemia; Saints Adalbert and Sigismund, signifying the faith of Charles IV and its importance for the Empire; the coats of arms of the crown lands with the Bohemian lion and the imperial eagle, representing the Emperor's claim to power; Charles IV himself; even King Wenceslaus.

To remind future generations of the atrocity committed on the Charles Bridge, the Hapsburg king Leopold I of Bohemia (reigned 1656–1705) had a monument to Johannes Nepomuk, as he was popularly known, erected in 1683. Saint John of Nepomuk has been revered throughout Europe since 1729. To this day, thousands of believers touch the head of the figure that Jean-Baptiste Mathey (c. 1630–95) portrayed falling into the river.

View of Charles Bridge

Charles Bridge, 515.76 meters long with 16 arches; named in 1870
after Charles IV under whom its construction began

Countless cannon-shots resounded
The château of Chenonceaux 1513–21

"Chenonceaux was entered along a lengthy avenue that was bounded by canals from which emerged sirens. Their song was answered by nymphs appearing from the woods... from the château countless cannon-shots resounded and fireworks exploded."

Royal celebration at the château of Chenonceaux, 1563

Phillibert Delorme
(1510–1570)

1547 builds the tomb of King Francis I at St. Denis
1550 supervises construction of château Anet, château Fontainebleau, and château Neuf at St. Germaine-en-Laye
1547–59 builds the chapel at Villers-Cotterets
1563 constructes extension of the Tuileries Palace in Paris

Jean Bullant
(1515–1578)

1530 builds château Écouen
1570 constructs additions to bridge at Chenonceaux

Catherine de Médicis spared no expense when, in 1563, she organized celebrations in honor of her son, King Charles IX of France (reigned 1560–74). A masked ball with a sumptuous banquet, a battue and a naval battle on the River Cher were held at the château of Chenonceaux.

King Henry II (reigned 1547–59), gifted the château of Chenonceaux to his favorite mistress, and Catherine de Médicis' rival, Diane de Poitiers (1499–1566). In 1556, the royal mistress commissioned master builder Philibert Delorme (1510–70) to add to the château a bridge of five arches from which—it is said—she would jump naked into the waters of the River Cher. Diane's carefree days came to an abrupt end, however, when Henry II was accidentally killed in a tournament—as prophesied by the king's personal physician, Nostradamus (1503–66): "On the field, the young lion will in single combat defeat the old lion. He will stab him in the eye...and the King will die a cruel death." Following the untimely death of her husband, Catherine de Médicis took revenge on her rival. Diane not only had to leave the court in 1559, but also lost Chenonceaux with the bridge she so dearly loved.

The Queen was not content with the cramped conditions of Chenonceaux's Renaissance rooms and its late Gothic chapel. As a venue for her lavish festivities, Catherine had two stories of arcades, including small terraces above the piers, added to Diane's bridge after 1570, probably to the design of Jean Bullant (1515–78). The two-storey bridge, seen in an engraving by the architect Jacques II Androuet du Cerceau (1550–1614) as a detail in a larger design, was to be the scene of numerous events in history: as a house of mourning for the French Queen Louise de Vaudémont (1553–1601), as a military hospital during World War I, and as a safe escape route to Vichy France during the time of the German occupation of northern France.

Diane de Poitiers, panel painting by François Clouet, 1571, National Gallery of Art, Washington D. C.

Catherine de Médicis in an anonymus contemporary painting

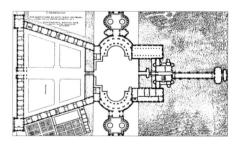

Engraving of Chenonceaux by Jacques II Androuet du Cerceau, 1607

The bridge at Chenonceaux as it appears today

A colorful rainbow
Mostar: "Stari most," the Old Bridge 1557–66

"With its two towers, this noble bridge reaches across the fast-flowing, dark river below, and with its proud arch it resembles a colorful rainbow."

Pascha Bajezidagi, late sixteenth century

**Süleyman I. Kanuni
(1494–1566)**

1494 born April 27 in Trabzon, the only son of Sultan Selim I. Yavuz (reigned 1512–20)
1520 elevated to caliph and sultan on September 21
1521 conquest of Belgrade
1522 occupation of Rhodes
1526 defeats the Hungarian King Ludwig II in the Battle of Mohacs; Hungary becomes a vassal state
1529–32 advances unsuccessfully against Vienna and the Habsburg Emperor Charles V (reigned 1519–56)
1550–56 commissions the famous master builder Kodza Mimar Sinan to build the Süleyman mosque in Istanbul; Sinan designs 330 other monumental structures in the Ottoman Empire
1566 dies September 6 in Sigeth

The sixteenth-century Ottoman writer and dervish Pasha Bajezidagi was not alone in admiring "Stari most," the Old Bridge at Mostar. Throughout history, the Old Bridge has been praised for its aesthetic qualities and daring form.

In 1557, the caliph and Sultan Süleyman I Kanuni (reigned 1520–66) commissioned the Ottoman architect Mimar Hajrudin to build a stone bridge across the Neretva, an ice-cold, emerald-green highland river. A pupil of the famous master builder Kodza Mimar Sinan, Hajrudin was to replace the wooden bridge that had become too small. Hajrudin chose the local, white limestone used to build all of Mostar's minarets as his building material. Hajrudin fortified the existing bridge by building stone arches on either side, using iron clamps to join the arch stones, and filling the void between wood and stone with molten lead. Hajrudin's initial attempt failed when the bridge collapsed as the scaffolding was being removed. As the scaffolding was being removed the second time, Hajrudin disappeared to the nearby village of Bijelo Polje for fear that the sultan would have him beheaded should the bridge collapse again. The master builder hid in a cemetery and, in tears, dug his own

grave. The daring arch, 28.7 meters long and 21 meters high, held. Hajrudin was saved and was richly rewarded by Sultan Süleyman I.

After the battle of Bijelo Polje in 1652, where 400 Mostar men fell and the Ottoman army defeated the Venetians, the "stone crescent," as the bridge was now called, was fortified by the addition of the Halebija defensive tower and the Tara munitions' towers. In 1664, the Ottoman traveller Evliya Çelebi noted in his diary that he had nowhere seen such a high bridge or such happy children as those who used the bridge to jump into the river below.

On November 9, 1993, the "stone crescent" and "colorful rainbow" was destroyed by Bosnian Croat artillery fire during the Bosnian civil war. In the words of the Croatian Franciscan Father Daniel, "The day the Old Bridge was destroyed was the day our town received a death sentence...When it collapsed, Mostar's heart was destroyed."

top left: The Old Bridge was destroyed in November 1993

Engraving of the bridge at Mostar

"Stari most," the old bridge leading to Mostar across the river Neretva in Herzegovina, 1557–66

On the Sea
Shanghai: The Zig-Zag Bridge 1559–77

"Shanghai is often the target of pirates. For that reason the coast is defended by means of significant fortifications and warships. The city takes its name from its proximity to the sea: Shanghai means 'on the sea'…The city is full of gardens… and officials, who on completion of their service, now enjoy their wealth as private persons."

Nicolas Trigault, *De christiana expeditione apud Sinas suscepta ab societate Jesu*, 1615

In 1609, on his journey through China as a missionary, the French traveller Nicolas Trigault (1577–1628) visited Shanghai. Still a town of no great significance, it nevertheless contained splendid parks, a spa resort and the retirement homes of numerous court officials. Most of the parks were closed to Trigault, as they were private gardens for the sole use of their owners and a circular wall shielded them from the inquisitive gazes of outsiders. This was true of the "Yu Yuan," known as the "Garden of Quiet Joy," where a wooden bridge zigzagged its way across a green lotus pond.

Between 1559–77, Pan Yunduan, the finance minister and governor of Suchuan province, had the "Yu Yuan" garden laid out for his parents' pleasure in their retirement. A dignitary under the Ming Emperors Chia-ching (reigned 1522–67), Lung-ch'ing (reigned 1567–72) and Wan-li (reigned 1573–1620), Pan Yunduan was determined that "Yu Yuan" would create a magnificent impression. He wanted his parents to be able to enjoy a different view across the lotus pond towards the garden as they crossed the "Bridge of the Nine Bends," as it was also called. The garden included thirty different landscapes, with meadows and hills, rocks and pavilions,

rest areas and paths. As it was longer than it needed to be to span the water, the Zig-Zag Bridge made both the lotus pond and the garden's relatively modest area of two hectares appear larger. This bridge had nothing in common with the traditional stone zig-zag bridges known in China since the fifth century, whose shape was developed to reduce and eliminate water resistance. By having users cross the bridge this way and that, Pan Yunduan was thinking more of the widespread principle of yin and yang, an awareness of the unity of opposites. Moreover, the design of the bridge was intended to prevent evil spirits from gaining access to the tea ceremony in the Huxing-Ting teahouse.

After the death of the last Ming Emperor Ch'ung-chen (reigned 1628–44), both the "Bridge of the Nine Bends" and its surrounding garden fell into disrepair, but reopened to the public in 1961 following extensive restoration work. European visitors to Shanghai now had their first opportunity to visit the "Garden of Quiet Joy" and to cross its wooden Zig-Zag Bridge, something that Nicolas Trigault in 1609 had found impossible to do.

Title page of Nicolas Trigault's *History of the Jesuit Mission in China*, University Library, Heidelberg, 1615

The "Bridge of the Nine Bends" on the left leads to the Tea House in the Yu Garden, print in *Reise der österreichischen Fregatte Novara*, volume 2, 1861

The Zig-Zag Bridge in Shanghai's
"Garden of Quiet Joy"

Bassano, at the foot of the Alps

Bassano del Grappa: Ponte degli Alpini sixteenth century

"At Bassano, an area at the foot of the Alps that divides Italy from Germany, I have constructed a wooden bridge across the river Brenta…it has posts which carry a roof serving as a form of loggia that lends beauty and practicality to the whole structure."

Andrea Palladio, *I quattro libri dell'architettura*, 1570

**Andrea Palladio
(1508–1580)**

1540–80 in Rome, his study of the writings of Vitruvius greatly influences his designs: dignity and austerity of architecture, clarity and scale of his proportions, their elegance and beauty give rise to the classical style of "Palladianism."
1549–80 constructs numerous churches, villas, townhouses and theaters, mainly in Vicenza, Venice, Lonedo and Udine
1554 designes for the Rialto Bridge in Venice
builds wooden bridge at Vicenza (destroyed)
1575 builds stone bridge across the river Guà at Montebello

Aesthetic appeal and practicality characterized the work of the famous architect Andrea Palladio (1508–80) when, in 1567, he was commissioned by the city fathers of Bassano del Grappa to build a wooden bridge. He alone was thought capable of designing a bridge that would both withstand the force of the raging river Brenta and be able to take the weight of the horses and carts carrying the town's commodities such as wool, silk and gold.

In his *I quattro libri dell'architettura* (Four Books on Architecture), Palladio described his solutions to the structural problems he faced: "At the spot where the bridge was to be built, the river measured 180 feet across. This distance was sub-divided into five equal sections…and four rows of piles were driven into the riverbed and beams were placed across…as the distance between the said rows is very great and because the longitudinal beams would have had difficulty in carrying particularly heavy loads, other beams were put in place that acted as brackets and which carried a part of the weight. In addition, other beams were attached to the piles in the riverbed. These beams were inclined towards each other and were connected with a further beam running lengthwise in the center. This arrangement of wood had the appearance of an arc…and gave the bridge a pleasant aspect, while giving it strength."

Just as he had demanded in his theory of the architecture of wooden bridges, Palladio expressed the belief that his Ponte degli Alpini would not be harmed either by the number of people and animals using the bridge or the weight of carts and guns or high water and flooding. Palladio was wrong, however, because in 1748 the Brenta swept away Bassano del Grappa's landmark bridge. Three years later, it had to be rebuilt to the design of Bartolomeo Ferracina, which incorporated modifications that strengthened Palladio's bridge. At the Battle of Bassano del Grappa on September 8, 1796, where Napoleon was victorious over the Hapsburg army, the bridge was able to meet the demands placed upon it.

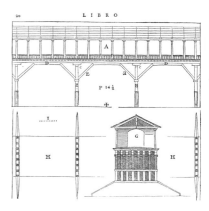

1567 design by Andrea Palladio for the Ponte degli Alpini, published in his *Four Books on Architecture* in 1570

Spanning the river Brenta, the Ponte degli Alpini was built by Andrea Palladio, 1567

The Heart of Paris
Pont Neuf: The New Bridge 1578–1607

"The Pont Neuf is to Paris what the heart is to the human body. The Bridge is the center of movement and circulation... it suffices to promenade there for an hour each day to meet the person one is seeking. Spies lie in wait there and if, at the end of a number of days, they have not seen their man, it is confirmation that he is far from Paris."

Louis-Sébastien Mercier, *Tableau de Paris*, 1782–88

Busy shop owners on the bridge: Nicolas and Jean-Baptiste Raguenet, *The Pont Neuf*, 1755, Musée Nissim de Camondo, Paris

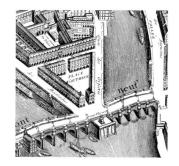

Pont Neuf and Place Dauphine on the Plan de Turgot, 1739. The equestrian statue of Henry IV is seen in the center of the bridge

The Pont Neuf was still largely free of today's level of hustle and bustle during dramatist Louis-Sébastien Mercier's day (1740–1814) but it was still the center of daily life in Paris. On its pavements, bootblacks and flower girls, teeth-pulling smiths and dog shearers offered their services alongside prostitutes. Singers and jugglers entertained the crowds. Mill wheels clattered beneath the bridge where bathers also sported in the river.

Since its completion, the Pont Neuf was a stage on which historic events were played out, tragic and festive alike. Think of the funeral procession of the murdered king Henry IV (1610); the mob tearing the limbs off the body of the hated minister Concino Concini (1617); and the triumphal processions of Emperor Napoleon I (reigned 1804–15). The reason why the Pont Neuf was chosen as the scene of such important events was simple: it was roomy and convenient. The shops and stalls that were found on other bridges in the city were absent here and so bridge users found their progress unimpeded.

The idea of a representative Seine crossing arose in 1577 when King Henry III of France (reigned 1574–89) was approached by Pierre Lhullier, the speaker of the city's traders, with a request to consider building a new bridge at the western end of the Île de la Cité. The king laid the foundation stone on May 31 of the following year. The bridge was designed by the architects Baptiste Androuet du Cerceau (c. 1545–90) and Métézeau Pierre Chambiges François des Illes (1534–1608). Guillaume Marchand Thibeau was in charge of construction. Together they created the longest Seine bridge in Paris with its unmistakable pier caps of round projections and grimacing faces. In 1588, religious strife brought an end to construction and only at the command of King Henry IV (reigned 1589–1610) did work on the project resume. Between the arches of the bridge at the tip of the Île de la Cité, a platform was created on which an equestrian statue of the Bourbon king was unveiled in 1614. Like the Pont Neuf, the statue was aligned with the Place Dauphine that Henry IV had built in honor of his son and heir Louis XIII (reigned 1610–43).

Still the oldest Seine crossing in Paris, the Pont Neuf, or "New Bridge," dominates the west of the Île de la Cité, 1578–1607

To serve the dignity of Venice
The Rialto Bridge across the Grand Canal 1588–91

"It would have looked splendid at the spot where it ought to have been built, namely in the center of one of Italy's largest and noblest towns…The Canal is exceptionally wide and the bridge would have been built where merchants from across the world meet to conduct their business."

Andrea Palladio, *I quattro libri dell'architettura,* 1570

**Antonio da Ponte
(1512–1597)**

1563 promoted by the magistrates of Venice to the office of "Proto al Sal," superintendent of public works
1579–82 manages extension of the Arsenal and, in Andrea Palladio's absence, supervises building work on the "Rendentore" church
1580–97 constructs the "Palazzo delle Prigioni," Venice's new prison
1588–91 reconstructs the Rialto Bridge

top left: The Rialto Bridge in an etching by Antonio Visentine, 1742. The etching is based on a painting by Canaletto

The Italian architect Andrea Palladio (1508–80) did not even mention Venice or the Grand Canal when he published his design for the Rialto Bridge in his *Four Books on Architecture.* In 1554, along with Jacopo Sansovino (1486–1570) and Giacomo Barozzi Vignola (1507–73), Palladio entered a competition to design the Rialto Bridge. He had already gained fame well beyond Italy as the architect of a number of villas, townhouses and churches in Vicenza and Venice. "To serve the dignity of the city," as he himself wrote, he planned three larger and two smaller arches, a four-columned loggia surmounted by the personification of Justice as well as recesses above supporting pillars for more statues. His design was inspired by the architecture of Roman antiquity—the very reason why his proposal was rejected. The competition judges felt that Palladio's representative architecture would be ill-suited to the demands of everyday life. Palladio's competitors, Sansovino and Vignola, saw their proposals rejected as well.

The moving wooden bridge of 1432 that allowed large spice-laden galleys to pass on their way to merchants' warehouses was still in use, accommodating twelve small shops on either side of the bridge which sold books, perfumes and confectionery. By 1587, however, the bridge's badly dilapidated state forced the Senate to organize a new competition. Unlike Vincenzo Scamozzi (1552–1616), who followed Palladio's design of 1554, the architect Antonio da Ponte (1512–97) was successful. In 1591, helped by his nephew Antonio Contino (1566–1600), da Ponte completed a bridge with a single arch spanning 27 meters, a design that revealed a hitherto unknown degree of daring, one that was interpreted by his contemporaries as a symbol of Venice's greatness. The bridge was high enough to allow the doge's state barge to pass beneath it and at the same time was shallow enough to accommodate 24 small shops beneath two rows of arcades rising towards its center.

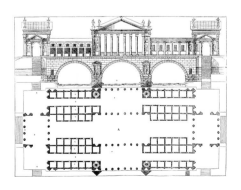

Wooden bascule bridge that was the predecessor of the Rialto Bridge, painting by Vittore Carpaccio, *The Miracle of the True Cross* (detail), c. 1494, Galleria dell' Academia, Venice

Design for the Rialto Bridge, drawing by Andrea Palladio, 1554; Museo Civico, Vicenza, Sheet D 25

Venice, the Rialto Bridge by Antonio da Ponte, a masterpiece of foundation work using
cofferdams and with radiating courses of abutment and spandrel masonry, 1588–91

The Pearl of Islam

Isfahan: Allahverdi Khan bridge 1599–1602

"From the shore of the Zayandeh river…we were afforded a view of three bridges, strange yet splendid buildings. The bridge that carried us across the river had 33 bottom-most arches; above each of them there rose three smaller ones. A covered walkway is provided with a surface of paving stones on the one level."

James Justinian Morier, 1808–16

Isfahan first came to prominence under Shah Abbas I the Great (reigned 1587–1629) in 1598, when he moved his court from Qazvin to Isfahan on the occasion of Persian New Year celebrations. At first, fate dealt Abbas one blow after another. His mother, elder brother and the grand vizier of his father, Shah Muhammad Khudabanda (reigned 1578–87) were murdered by power-hungry Qizilbash emirs. The governor of Mashhad established Abbas in opposition to his father in 1581. When his father abdicated six years later, all responsibility fell to Abbas. Through diplomatic and military skill, he defeated the Ottomans and brought peace to the Safavid empire. Trade, science and the arts flourished—mainly in the royal residence of Isfahan, home to 600,000 of his subjects. On account of its splendid buildings, among them the Maidan square, the Shaikh Lutfallah and Shah mosque, the Ali Qapu palace or the gardens of Chahar Bagh, contemporaries described Isfahan as "half the world."

Much admired by Morier in 1808, the three bridges across the river Zayandeh added to Isfahan's reputation in earlier centuries. Besides the Canal Bridge and the Khaju Bridge, built later under Shah Abbas II (reigned

1642–67), it was mainly the Si-O-Se Pol Bridge with its thirty-three arches that made an impression on Morier. Shah Abbas I the Great commanded his celebrated general and minister Allahverdi Khan to construct it between 1599 and 1602. With its carriageway for beasts of burden and wagons, it not only linked the upper and lower halves of Isfahan's principal thoroughfare, but also served as a meeting place for the local nobility. Shah Abbas I the Great wanted everyone approaching the Allahverdi Khan bridge, as it was later called, to know that they were about to enter the "Pearl of Islam," the foremost city of the Safavid empire, Isfahan. James Justinian Morier visited Persia when Isfahan's golden age had long since passed. In 1722, the Ghilzay Afghans under their leader Mahmud laid waste to the city.

Coupe du Pont Alyverdy-Chan.

Views of the Allahverdi Khan Bridge, from *Historic Architecture* by Johann Bernhard Fischer von Erlach

At 295 meters long, 13.75 meters wide and with a 5.6 meter arch span, the Allahverdi Khan bridge is the longest crossing over the river Zayandeh

A shudder seized me

Venice: The Bridge of Sighs 1600

"Messer Grande, the highest of the Republic's civil officers, and I, Giacomo Casanova, crossed a steep, enclosed bridge that connects the prison with the Doge's palace across the Rio di Palazzo canal. On the other side of the bridge, we walked along a gallery and…he handed me over to the prison warder."

Casanova, Girolamo-Giacomo, *Histoire de ma Fuite*, 1787

Antonio di Bernardino Contino (1566–1600)

1566 born in Lugnano, possibly a nephew of Antonio da Ponte
trains as a stonemason and architect
constructs the Campanile of S. Giorgio dei Greci in Venice
assistant to Antonio da Ponte during construction of the Rialto Bridge
1591 elected "proto" of the ducal palace
1597 elected "proto" of the "Ufficio del Sale"
1597 completion, with the help of his uncle Tomaso di Francesco Contino, of the New Prison begun by Antonio da Ponte
1600 Bridge of Sighs built
1600 dies mid-July in Venice

Ladies' man Girolamo-Giacomo Casanova (1725–98) later recalled the "Ponte dei Sospiri," the Bridge of Sighs, as a steep, enclosed bridge when he was arrested during the night of July 25 to 26, 1755, and sentenced to five years' imprisonment for occultism, freemasonry, sexual offences and blasphemy. When Casanova was led to jail past the bridge's plain inside walls, he could see nothing of its splendid facade. Nor would it have interested him for his thoughts were focused on his successful escape from Venice's "Piombi" prison on November 1, 1756.

In 1600, the stonemason and architect Antonio di Bernardino Contino (1566–1600) built the Bridge of Sighs to connect the Doge's Palace, where criminals were sentenced, and the "Piombi" prison. He added wall decorations and scrolls, including a personification of Justice in the segmental arch to establish a link between the bridge and the Venetian judiciary, and the coat of arms affixed below to reference the representative of the Republic of Venice, the Doge Marino Grimani (in office 1595–1605). The bridge was a continuation of the work on the new prison that Antonio, with his uncle Tomaso di Francesco Contino, completed after 1597.

Even after the Republic of Venice lost its independence in 1797, the bridge across the Rio di Palazzo continued to engage the imagination of poets. In 1817, the English Romantic poet Lord Byron (1788–1824) named it the "Arch of Sighs." Two years later, the Austrian dramatist Franz Grillparzer (1791–1872) recorded in his diary, "Sailing past the State Prison in the moonlight, through the shadows which these massive buildings furtively cast one upon the other, shadows broken by the occasional flash of light, suddenly there soared above me the Ponte dei Sospiri across which political prisoners were once taken to their death. A shudder seized me. These bygones and deceased, persecutors and persecuted, murderers and their victims all seemed to rise up before me, their heads covered."

Portrait of Girolamo-Giacomo Casanova by Pietro Longhi (1702–85)

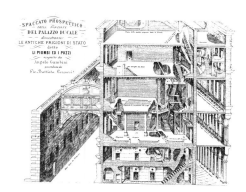

The Bridge of Sighs connecting the Doge's Palace and the New Prison

The Bridge of Sighs, built in 1600 by Antonio Contino

In the Garden of Perfect Clarity
Beijing: The Jade Belt Bridge 1751–64

"My grandfather, the Emperor, taught me that the meaning of all things on earth lies in their perfection and clarity. The significance of yuan lies in its perfection, that of ming in its clarity and brilliance. For that reason he chose to call the park the 'Garden of Perfect Clarity,' for these two qualities are among a ruler's greatest virtues."

Emperor Ch'ien-lung on the splendid garden at his Summer Palace, *c.* 1736

Emperor Ch'ien-lung (1711–1799)

1711 born the fourth son of Emperor Yin-jeng (reigned 1723–36)
1736 follows his father Yin-jeng (reigned 1723–36) as Emperor of China; under his reign, China reaches the peak of its economic and cultural development
1751–64 Yuanming Yuan becomes the imperial summer residence
1754 completion of the Temple of Heaven in Beijing
1755–60 construction of Puning-Si, the Temple of Universal Peace
1758–59 suppression of the first Mongol rebellion in Turkestan
1767–69 Burma becomes a tributary state
1781–82 suppression of the second Mongol rebellion
1783 reconstruction of the central hall of the Imperial Academy in Beijing
1792 suppression of the Ghurkha rebellion
completion of the Temple of White Clouds in Peking
1796 transfer of power to Yung-yen (reigned 1796–1820)

The Manchu Emperor Ch'ien-lung (reigned 1736–96) was filled with deep respect when he wrote about his grandfather K'ang-hsi (reigned 1661–1722) and his Yuanming garden. The garden held special meaning for the Emperor because his father, Yung-cheng (reigned 1723–36), died there following a riding accident. On his deathbed, he entrusted the empire and his much-loved garden to the care of his son. After three years of mourning, Ch'ien-lung made it his main residence. After 1751, he tripled the size of the park because he planned to gift it to his mother on her sixtieth birthday. Work on the garden was delayed until 1764, however, and it was no longer possible to make a gift of it. Ch'ien-lung thus chose the "Garden of Perfect Clarity" as his summer residence. Among other things, it contained the Seventeen Arch Bridge (shiqi-konqiao) that was renowned for the 500 stone lions that lined its parapet and that led to Nahu island with its Dragon King Temple.

The garden contained the Jade Belt Bridge, also known as the "Camel Back Bridge" due to its steep arch. Ch'ien-lung had the garden designed in a way that accentuated the bridge's decorative role and that ensured its noble, gleaming white masonry was reflected in the dark water

of Kunming lake. The image was to suggest a union between the watery-green curve of the river dragon and that of the moon. The great height of the arch also allowed the emperor to sail under it in his ceremonial barge and to view the lake and surrounding garden and palaces. From the vantage point of the Jade Belt Bridge, he could grasp the function of the Yuanming Garden as a mirror image of the whole of the Manchu kingdom—that is if he was not busy tilling the fields, tending his vegetable plots or seeing to his silkworms. This was how he came to know the hard work done by his peasant subjects and to appreciate the sunshine and the rain. Ch'ien-lung recorded enthusiastically: "As the heir of my father and grandfather and their principle of simplicity, I am happy to live in this place that is protected by heaven and blessed by the earth." The clouds in this little bit of heaven turned ominously dark in 1860 and 1900 when the Yuanming Garden was laid waste in the second Opium War and Boxer Rebellion, respectively. A former imperial concubine, Tz'u-hsi (1835–1908), restored the devastated park surrounding the Jade Belt Bridge in 1888/95 and 1903/08 and named it Yíhé Yuan, the "Garden of Harmonious Union."

Emperor Ch'ien-lung on the Imperial Throne; gouache by William Alexander (1767–1816), British Museum, London, 1795

The Seventeen Arch Bridge with its 500 stone lions

Its distinctive shape earned the Jade Belt Bridge in the Emperor's Summer Palace at Beijing the nickname of "Camel Back"

A Sand-Cast Bridge

The Iron Bridge, Coalbrookdale 1776–79

"Sand-casting was the method used to manufacture everything. Once scaffolding had been erected, each section of the ribs was raised to a certain height with the aid of strong ropes and chains and then lowered again until the upper ends reached the mid-point. The main sections were erected over three months, with no accidents either in the foundry or among the men on site; neither was traffic on the river impeded."

Riedel on the manufacture of the Iron Bridge at Coalbrookdale, 1797

Thomas Farnolls Pritchard (1723–1777)

1723 born in Shrewsbury, the son of a carpenter
1746 works as a copperplate engraver
1749 builds St. Julian's Church
1750s works as a property assessor, renovates Dothhill Park House, Shrewsbury School and the Hosier Almshouses
1770 leaves Shrewsbury and takes lodgings in Eyton-on-Severn where he designs bridges (Stourport, English) and safeguards the bridge over the River Teme near Downton Castle
1772 drafts plans for the renovation of Powis and Downton Castles
1777 dies December 21

top left: William Williams' view of the Iron Bridge, 1780, The Iron Bridge Gorge Museum in Telford

The opening of the world's first cast-iron bridge near Coalbrookdale on the river Severn on New Year's Day, 1781 was celebrated as a milestone in the Industrial Revolution. It was a technology that had long been regarded with suspicion. The English engineer John Smeaton (1724–92) commented, "When I first used cast iron for certain purposes twenty-seven years ago, everybody cried 'How do you expect brittle iron to last when the strongest wood does not?'"

Nonetheless, such doubts did not deter ironmaster Abraham Darby II (1711–63), whose Coalbrookdale Works was the first company to succeed in smelting iron ore with coke, although he died before realizing his plan to build a cast-iron bridge. Despite the high financial risk involved, Abraham Darby III (1750–91) took up his father's plan to build a cast-iron bridge across the River Severn near Coalbrookdale as a replacement for a permanently overcrowded ferry. Darby chose a design by the architect Thomas Pritchard (1723–77) from Shrewsbury, a man who had worked on cast-iron bridges since 1773. He recognized that iron bridges had greater compressive strength and resistance than wood or masonry structures, and his design included a complete arch of cast-iron members between brick abutments. Parliamentary approval for the bridge was given on March 25, 1776. John Wilkinson, the "great Staffordshire ironmaster" (1728–1808), was involved in its execution. It was through Wilkinson, in fact, that Abraham Darby III met Pritchard. Casting of the five parallel ribs that formed the semi-circular compressive arch had just begun. They were mounted on cast-iron bed-plates so as to distribute their weight across the foundation.

At the end of July 1779, erection of the bridge began and lasted over seventeen weeks. The bridge's cast iron pieces were connected by mortise joints and pegged dovetails. The final additions were the railing and the road surface of clay and iron slag. The result was a success and gave rise to an industrial town that took the name Ironbridge. Pritchard did not live to see the realization of his design. Dogged by illness for the last year, he died on December 21, 1777.

The first imitations appeared in Raincy near Paris (1788), in the grounds of the palace at Wörlitz in Saxony (1791) and at Laasan, Lower Saxony (1796), providing adequate proof that by the end of the century, the Continent, too, appreciated the quality and beauty of cast-iron bridges.

The toll depended on what mode of transportation one used

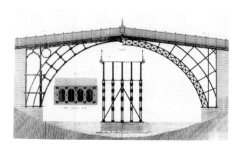

Illustration of the bridge before it was modified in 1821, steel-plate engraving after Pritchard, 1779

The Iron Bridge near Coalbrookdale, the world's first cast-iron bridge; it weighed 378.5 tons and had a span of 30.6 meters. Its masonry abutments were replaced in 1821 by iron arches

The most wonderful feat
Bristol: The Clifton Suspension Bridge 1831–64

"Of all the wonderful feats I have performed, since I have been in this part of the world, I think yesterday I performed the most wonderful. I produced unanimity among fifteen men who were all quarrelling about that most ticklish subject—taste."

Isambard Kingdom Brunel in a letter to
his brother-in-law Benjamin Hawes, 1829

**Isambard Kingdom Brunel
(1806–1859)**

1806 born April 9 in Portsmouth, the son of engineer Sir Marc Isambard Brunel (1769–1849)
1824–43 construction of the Thames Tunnel built by his father's company
1833–58 builds the Great Western railway line with a total of 64 viaducts, among them the Gover Viaduct (210 m), the St. Germans Viaduct (288 m) and the St. Pinnock Viaduct (193 m)
1838 construction of the Maidenhead Bridge
1841–45 builds the Hungerford Suspension Bridge across the Thames
1853–59 completes the Royal Albert Bridge across the river Tamar to link Saltash and Plymouth
1854 completion of Paddington Station, London
1859 dies September 15 in the City of Westminster

Isambard Kingdom Brunel (1806–59) was a mere twenty-three years old when he convinced the Clifton Bridge Company, whose committee was headed by the renowned Scottish civil engineer Thomas Telford (1757–1834), of the merits of his design for a suspension bridge across the Avon Gorge. At the first meeting he presented four sketches of suspension bridges that envisaged far more daring constructions than Telford's bridge over the Menai Straits (1819–26). During a second round of discussions, Brunel presented to the Bridge Company a plan for a shortened span of only 183 meters—and won the contract.

Although a young Brunel began the project in 1831, lack of finance delayed the laying of the foundation stones for the towers and abutments until August 27, 1836. By this time, however, he had been entrusted with a far greater task, the construction of the Great Western railway between Plymouth and Truro for which alone 64 viaducts had to be built. From his position of strength, this increasingly famous timber engineer was able to reject the Clifton Bridge Company's suggestion of suspending the wooden deck from wire cables. Instead, he chose two pairs of iron chains. For lack of finance, he was

again instructed to stop work in 1842. Seven years later, the chain iron for the bridge was sold and completion of the project seemed a distant prospect. When Brunel died in the City of Westminster on September 15, 1859, an incomplete structure still straddled the Avon Gorge. Both towers, reminiscent of ancient Egyptian architecture, stood abandoned in the landscape.

Things changed only when John Hawkshaw (1812–91) and others among Brunel's colleagues at the Institution of Civil Engineers founded a company to complete the Clifton Suspension Bridge between 1862 and 1864. By chance, Brunel's Hungerford Suspension Bridge was then being demolished and its chains were bought for use at Clifton. As they were lighter than those Brunel had planned to use on the Clifton Suspension Bridge, Hawkshaw and William Henry Barlow (1812–1901) added a third chain at either side. The deck, too, was realized in modified form, narrower, higher and made of wrought iron. Brunel had planned to use wood because he had no faith in cast iron. Nevertheless, when the Clifton Suspension Bridge was completed, it was essentially to Brunel's design, one whose creator had no reason to believe would ever be completed.

The Clifton Suspension Bridge at Bristol, color lithograph from a private collection

The Clifton Suspension Bridge spans the Avon Gorge at a height of 74.7 meters and is 214 meters long

Defected by canon balls

Budapest: the Széchenyi Chain Bridge 1840–49

"The bridge has suffered comparatively little by the springing of a mine and the piers are in several places defected by canon balls. Thirty-two of the finest houses or rather palaces of Pesth are destroyed by fire and several hundred others are more or less damaged by shot and shells."

Adam Clark in a letter to his parents, Pest, May 27, 1849

William Tierney Clark
(1783–1852)

1783 born August 23 in Bristol
1808 hired by the Scottish engineer John Rennie to work at his Albion Ironworks at Blackfriars, London
1811 Begins as an engineer for Middlesex Waterworks Company
1824–27 construction of Hammersmith Bridge across the Thames
1832 completion of the Marlow suspension bridge across the Thames
1852 dies in September, buried in St. Paul's, Hammersmith

Adam Clark (1811–66), the Scottish engineer leading the construction of Budapest's Chain Bridge, tried to prevent the Austrian commanding officer in Buda, Colonel Allnoch, from blowing up the unfinished bridge. Unsuccessful, Clark watched as Allnoch used a cigar to ignite 400 kg of gunpowder on the deck of the bridge on May 21, 1849. Fortunately for Clark, the bridge held; Allnoch, however, was torn to bits.

Since the brutal murder of the Austrian envoy, Count Lamberg, in 1848, Hungary had been in open revolt against the Austrians. The freedom fighter Lajos Kossuth (1802–94) was attempting to rid his country of its Austrian rulers. The bridge, only provisionally finished, passed its first test in 1848 when 70,000 soldiers and 300 pieces of military equipment used it to cross the Danube. Despite several victories, the Hungarian general Artúr Görgey (1818–1916) had to yield to superior Austrian and Russian forces. He capitulated at Világos following the cruel suppression of his revolt by prince Windisch-Graetz.

Besides use during times of war, the Chain Bridge was formally opened on November 20, 1849. By linking the towns of Buda and Pest, the bridge symbolized the unity of the Hungarian nation. The minister of transport, Count István Széchenyi (1791–1860), who commissioned the bridge, was not among the guests. He was under house arrest for the active part he had taken in the Hungarian revolution.

And yet everything had started so well! In August 1832, he and Count György Andrássy had set off enthusiastically for England where they met the engineer William Tierney Clark (1783–1852) and visited his Hammersmith Bridge in London. The decision was made to replace the wobbly, wooden pontoon bridge across the Danube with a suspension bridge. With the help of an association that Széchenyi founded, banks were persuaded to finance the project and the Englishman William Tierney Clark was hired as its engineer. After 1840, with the help of Adam Clark (no relation), he began work. Despite structural problems, the Széchenyi Bridge stood the test of time until January 18, 1944 when the retreating German army succeeded in blowing it up. The Hungarian poet and novelist Gyula Illyés wrote, "The saddest sight was the bridges between the two towns, just lying there, one after the other, like slaughtered animals, in ruins, innocent."

View of the bridge from Pest, *c.* 1900. The carriageway is still in its original condition and has fine wood and iron stiffeners

Széchenyi Chain Bridge, 380 meters long, William Tierney and Adam Clark, 1840–49; rebuilt in 1949 to mark its centenary

The Eighth Wonder of the World
The Göltzsch Valley Bridge in the Vogtland 1846–51

Look upon this masterpiece/The eighth wonder of the world, the Göltzsch Valley Bridge!
This bridge that arches up to the blue of the sky/Is dedicated to its valiant builders.

Declaimed as the keystone was set in place,
September 14, 1850

Johann Andreas Schubert
(1808–1870)

1808 born March 19 in the Vogt-
land, the son of an impoverished
farmer
1824–28 studies at the School of
Building at Dresden's Art School
1828 assistant to the mathematics
professor at Dresden's Technical
College
1832 becomes professor at
Dresden's Technical College
1836 sets up a mechanical engineering company to manufacture
efficient steam boilers
1837–39 builds first German steam
locomotive, *Saxonia*, first Saxon
Elbe steamer, *Königin Maria*
1845 structural analyses for the
Göltzsch and Elster Valley bridges
1852 member of State Examination
Commission for Engineers
1859 dubbed a knight of the Royal
Saxon Order of Public Service
1870 dies October 6 in Dresden

The guests attending the ceremony marking the completion of the Göltzsch Valley Bridge in 1850 regarded the
construction as the eighth wonder of the world—and
not without reason. At 78 meters high and 574 meters
long, the bridge had been built to symbolize a successful
process of industrialization. King Frederick August II of
Saxony (reigned 1836–54) set the last stone and with
three hammer blows thanked all those whose physical
strength and intellectual power had contributed to the
project. In his speech, Prince Albert (1828–1902) in particular stressed the name of foreman Robert Wilke as the
creator of the bridge. No mention was made of its structural engineer, Professor Johann Andreas Schubert
(1808–70), who had fallen out of favor, having two years
earlier expressed his support for democratic principles in
an "Open Declaration" to the German National Conven-
tion at Frankfurt. Initially, Schubert enjoyed the greatest
confidence and in 1845 was installed as the chairman of
a committee that was to judge designs submitted for an
international competition. He was, after all, the man who
built the first German steam locomotive, the Saxonia.

King Ludwig I of Bavaria (reigned 1825–48) signed a
treaty in 1841 with the King of Saxony that would allow
the construction of a railway line from Nuremberg to
Leipzig, an undertaking that would require engineers to
span the deep valleys of the Göltzsch and Elster rivers.
None of the 81 submissions passed the scrutiny of the
judges and so the Interior Ministry of Saxony assigned
the task of structural analysis to Professor Schubert himself. He favored a solid construction of granite and brick.
Taking the view that a pier construction with round arches would be both the cheapest and sturdiest option,
Schubert designed a four-tier viaduct across the Göltzsch.

The first train crossed the Göltzsch Valley Bridge on
a rainy July 15, 1851, a much-anticipated event that was
attended by Prince Albert and other prominent figures
from Saxony. Yet again, no invitation had been extended
to Schubert. The annals of the Saxon Engineering Association record the following comment: "Whenever musi-
cians play a waltz, the program mentions the composer's
name. Whenever a steam locomotive thunders across the
tallest of bridges at dead of night in complete safety—
bridges built under inconceivable hardship and danger—
only too often in a short space of time no one round
about can recall the name of the man who designed
them."

The Göltzsch Valley Bridge under construction,
lithograph by W. Bässler, 1850

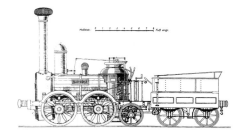

Saxonia and tender, the first usable, all-German
steam locomotive, built by Johann Andreas
Schubert, 1838

The Göltzsch Valley Bridge, once the world's highest
railway bridge, 1846–51

Beauty unsurpassable

James Eads' St. Louis Bridge   1867–74

"I have haunted the river every night lately, where I could get a look at the bridge by moonlight. It is indeed a structure of perfection and beauty unsurpassable, and I never tire of it."

Walt Whitman, 1879

**James Buchanan Eads
(1820–1887)**

1820   born May 23 in Lawrence-burg, Indiana, and named after the American president James Buchanan
1833   moves to St. Louis where he trains as an engineer
1862   builds a number of ironclad gunboats for the use of the Union in the American Civil War: *Cairo, Carondelet, Cincinnati, Louisville, Mound City, Pittsburg* and *St. Louis*
1867–74   Eads' Bridge at St. Louis built
1879   completion of the New Orleans shipping channel
1884   first American to be awarded the Albert Medal of the Royal Society of Arts in London for his design of the St. Louis Bridge
1887   dies March 8 in Nassau, Bahamas
1920   first engineer to be included in the American Hall of Fame

The American poet Walt Whitman (1819–92) was not alone in praising James Eads' bridge at St. Louis. "Sensational" and "groundbreaking" was the view of the local press and other American newspapers; even European journals enthused about the bridge. The residents of St. Louis still recalled Charles Ellet's (1810–62) attempt in 1840 to throw a bridge across the Mississippi. It failed on account of the river's width (460 m), its strong currents, muddy bottom and winter ice. It was all the more surprising when a new bridge company began to raise funds after 1865 for what appeared to be a lost cause and hired James Buchanan Eads (1820–87) as engineer-in-chief.

Despite strong opposition from riverboat owners, Eads started work in August 1867 on the west abutment. His plan was to build a steel bridge with three large spans measuring 153 m, 158.5 m and 153 m that would still allow paddle steamers to ply the Mississippi. Atop the spans he planned to build two storeys, an upper one for carriages, wagons and pedestrians and a lower one for modern steam locomotives. As construction of the piers proved to be particularly difficult, Eads decided to make use of pneumatic caissons that he had seen used during a trip to France. The piers and the east

abutment were founded using iron-clad timber caissons. Eads developed a pump that was used to remove the vast amounts of sand and mud from the compressed air chambers. Following construction work on the East Pier, twelve men died from "compressed air disease" ("the bends"). Despite such difficulties, Eads was able to complete the piers in 1871.

On May 23, 1874 his assistant wrote enthusiastically, "The first users of the completed bridge were not wagons or locomotives, but pedestrians. St. Louisians paid a nickel per person today for the right to be among the first to walk across the Mississippi River. The formal dedication of the St. Louis Bridge took place today, July 4, 1874—complete with a 14-mile-long parade, dozens of hooting steamboats and $10,000 worth of firework. Exciting! A plaque mounted on the structure declared: 'The Mississippi discovered by Marquette, 1673; spanned by Captain Eads, 1874.' When they asked Captain Eads if he is relieved that his bridge is a success, he claimed 'he felt no relief because I had felt no anxiety on the subject.' He predicted 'yon graceful forms of stone and steel' would endure as long as pyramids.' I'm sure he'll be proven right."

Upon completion, the St. Louis Bridge had the longest steel span in the world

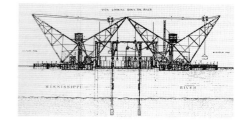

Schematic drawing of the caisson-sinking apparatus

James Eads' 480-odd-meter-long bridge across the Mississippi was
regarded as a wonder of the world by its contemporaries

The harp of death, the strange winged creature
New York: Brooklyn Bridge 1869–83

"The Brooklyn Bridge is one of the mechanical wonders of the world, one of the greatest and most characteristic of monuments of the nineteenth century...The approaches themselves are greatly impressive... that through them we get what is to be got nowhere else in our rectangular city, glimpses and 'bits' of buildings."

Montgomery Schuyler, "The Bridge as a Monument,"
May 26, 1883

**Johann August Roebling
(1806–1869)**

1806 born June 12 in Mühlhausen, Thuringia
1845 construction of Pittsburgh's Allegheny Bridge
1851–55 builds Niagara Falls Bridge
1866 Cincinnati Bridge opened, its span of 322 meters setting a new world record
1869 dies July 22 in Brooklyn

Construction of the Brooklyn Bridge, Illustration, 1881

In an article entitled "The Bridge as a Monument," Montgomery Schuyler paid tribute to the bridge that linked Brooklyn and Manhattan Island. Its span of 486 meters made it the world's longest suspension bridge. Incorporating peaked arches and modelled on the skyscrapers then found in the city, two 90-meter-high granite towers supported the deck by means of steel cables and an intricate web of radiating stays. The bridge had room for a central, elevated walkway, railway lines and a road. The moment had been a long time coming. New York's city fathers had decided to build a suspension bridge as early as 1857. Ten years later, German engineer Johann August Roebling (1806–69) won the design competition on the strength of his international fame. Before he started work on the foundations for the piers, however, the "harp of death" cast its shadow upon the project. During surveying work, Roebling became trapped between two survey vessels and was seriously injured; he died of his injuries three weeks later. His eldest son Washington August Roebling (1837–1926) took over as chief engineer. His biggest problem was positioning the pneumatic caissons in water up to 23.8 meters deep. News of the necessity for slow decompression had not yet reached New York and Roebling himself collapsed one day in 1872 upon leaving the caisson, a victim of caisson disease (the bends). Partially paralysed, deaf and with deteriorating eyesight, he now depended on his wife Emily as his messenger and colleague. Confined to his home in Columbia Heights, he used binoculars to supervise the progress of construction work, and in this way saw the project completed.

The Roeblings' Brooklyn Bridge not only impressed engineers and technologists. Regarded as the Eighth Wonder of the World, it also inspired writers such as Henry Miller (1891–1980), who was profoundly moved by it, and wrote, "Walking back and forth over the Brooklyn Bridge everything became crystal clear to me. Once I cleared the tower and felt definitely poised above the river the whole past would click. It held as long as I remained over the water, as long as I looked down into the inky swirl and saw all things upside down. It was only in moments of extreme anguish that I took to the bridge, when, as we say, it seemed that all was lost. Time and again all was lost, irrevocably so. The bridge was the harp of death, the strange winged creature without an eye which held me suspended between the two shores."

An impressive backdrop: the Brooklyn Bridge across the East River

The shape of a crescent moon
Porto's Maria Pia Bridge and Luís I Bridge 1876–77

"The arch of the Maria Pia Bridge, of particular appearance, rests upon simple abutments… Its height increases progressively in such a way that the bridge resembles a crescent moon. Its shape is especially convenient because it can withstand asymmetric forces and… permits extreme heights."

Théophile Seyrig on the Maria Pia Bridge, 1878

**Gustave Eiffel
(1832–1923)**

1832 born December 15 in Dijon, the son of Jean René Boenick-hausen, who adopted the name of his birthplace (northern Eiffel)
1850 studies at Paris Polytechnic
1856 company secretary and engineer's agent for the coach and steam engine builder Charles Nepveu
1867–85 his company builds bridges in France, Spain, Portugal, Romania, Hungary, Algeria, Peru and Indo-China
1889 the Eiffel Tower opens for the Paris International Exposition
1923 dies December 28 in Paris

Engineer Théophile Seyrig (1844–1923) spoke of the daring arch of the Maria Pia Bridge in the Portuguese port city of Porto in his 1878 book *Le pont du Douro à Porto*. Only Seyrig, the bridge's designer, and Gustave Eiffel (1832–1923), understood the unique structural challenges facing the railway bridge.

The depth both of the valley and the river Douro itself prevented the erection of a central pier or multiple pylons. The bridge's overall length of almost 353 meters therefore had to be spanned by a parabolic arch 61.2 meters high. Anchored in abutments at the foot of the valley walls, the centring for the steel arch had to be erected from either side of the Douro's steep riverbank. The precision with which both sections were assembled ensured that they joined exactly. When, on November 4, 1877, the 1,600-ton bridge was ceremonially named in honour of the reigning Portuguese Queen and opened to railway traffic, its span of 160 meters was one-and-a-half meters greater than that of the central arch of James Buchanan Eads' (1820–87) St. Louis Bridge across the Mississippi. Portugal's railway operators contracted Eiffel's Paris-based company to build the Douro Bridge, instead of a Portuguese firm, because of the recognition

that Gustave Eiffel enjoyed internationally. Moreover, his company had built viaducts along the railway lines from Poitiers to Limoges (1867), from Latour to Millau and from Brive to Tulle (1870). He had even built bridges as far afield as Romania, on the railway line between Jassy and Ungheni, and at La Oroya in Peru (1872).

Four years after the Maria Pia Bridge opened, Porto's city fathers, so convinced of Eiffel's pioneering engineering achievement and aesthetic responsiveness, commissioned Eiffel and his engineer, Théophile Seyrig, to build an even larger bridge, in honour of King Luís I of Portugal (reigned 1861–89), for vehicles and pedestrians. This black two-tier bridge had a span of 172 meters and, along with the Maria Pia Bridge, became a symbol of Porto. Because of these two magnificent bridges, "Bridges to the Future" was chosen as the city's motto during its term as European City of Culture in 2001.

Named in honor of the then queen of Portugal: Maria Pia Bridge across the Douro near Porto, 1877

Eiffel's bridges have become the symbols of Porto. The Dom Luís I Bridge is shown here

Strength and Stability
The Forth Bridge 1882–89

Twas about seven o'clock at night/And the wind it blew with all its might,
And the rain came pouring down/And the dark clouds seem'd to frown,
And the Demon of the air seem'd to say—I'll blow down the Bridge of Tay.'

William McGonagall

John Fowler
(1817–1898)

1817 born July 15 in Wadsley
near Sheffield
designer of St. Enoch's Station,
Glasgow
1860 designs Pimlico Bridge, the
first railway crossing of the Thames
1861 co-designer of Victoria
Station
1864 builds the Albert-Edward
Bridge at Coolbrookdale, Shropshire
1890 knighted for his work on
the Forth Bridge
1898 dies November 20 in
Bournemouth, Hampshire

Benjamin Baker
(1840–1907)

1840 born March 31 in Keyford,
Somerset
1879 transports Cleopatra's Needle
from Egypt and installs it by
the River Thames
1890 knighted for his work on
the Forth Bridge
designer of the first Hudson
River Tunnel
1907 dies May 19 in Berkshire

The terrible events at Dundee on the night of the December 28, 1879 long haunted the world. On a stormy winter's night, the bridge collapsed just as a train was crossing it. All seventy-five of the passengers and crew were killed as the train and bridge tumbled into the water of the Firth of Tay.

Only three years before, the railway-bridge had been hailed as a "triumph of engineering" that earned its designer, Thomas Bouch (1822–90), a knighthood and secured for him a contract to build the Forth Bridge. With his career in ruins, Bouch's design was immediately rejected and the project was entrusted to John Fowler (1817–98) and Benjamin Baker (1840–1907). With the Tay Bridge disaster ever present in their minds, the two engineers chose to construct a steel cantilever bridge that would be five times as strong as Bouch's design. According to Baker, the bridge was to epitomize strength and stability.

With the Forth reaching depths of 220 feet, they realized that it was impractical to build multiple piers to carry the many truss spans. Only the small rocky outcrop of Inchgarvie in mid-stream allowed the construction of a central pier of four circular caissons to support the bridge; another set of four circular caissons was constructed on the North and South Queensferry sides. The approaches to the bridge proper were formed by masonry viaducts at whose ends stood masonry piers to which the bridge's outer cantilever arms were tied. The two 350-feet suspended spans were built out over the Forth until they joined above the mid-channel point. On its completion, the Forth Bridge entered the record books for the height, length and depth of the cantilevers and the 55 thousand tons of steel consumed. It also cost 57 men their lives.

In 1889, the Scottish architect Rowand Anderson commented, "The designing of machinery has now reached such a high standard of excellence in function, form and expression that one is justified in saying that these things are entitled to rank as works of art as much as a painting, a piece of sculpture, or a building." The last rivet was ceremoniously driven home on March 4, 1890 by the Prince of Wales in the presence of Benjamin Baker and Gustave Eiffel, among others. Contemporary voices were critical, however. The *Times* reported that very few would find the bridge pleasing in appearance, and the designer and craftsman William Morris even described it as "the supremest specimen of all ugliness."

Model of the cantilever bridge with the Japanese assistant Kaichi Wantanabe in the center

The Forth Bridge as shown in *Illustration*, 8.3.1890

The world's greatest cantilever bridge, the Forth Bridge, has an overall length of 1.5 miles (2.46 km) and two clear spans of 1,710 feet

Stockbroker in armor

London: Tower Bridge 1886–94

*"The Tower Bridge is the most monstrous and preposterous architectural sham
that we have ever known."*

The Builder, London, June 30, 1894

**Sir Horace Jones
(1819–1887)**

1874–76 designs London's
Billingsgate Market as realized by
John Mowlem
1880–86 designs extension
to the London Guildhall School
of Music and Drama
1886 designs Tower Bridge with
the help of his assistant George
Stephenson
1887 dies while Tower Bridge
is still under construction

**Sir John Wolfe-Barry
(1836–1918)**

1836 born the son of Sir Charles
Barry (1795–1860), chief architect
of the Houses of Parliament
1886–94 works on the construction of Tower Bridge
1890s designs pumping station,
Regent's Canal Dock
1899–1903 designs and builds
Kew Bridge together with Cuthbert
Brereton
1900–07 serves as chairman
of Sir John Pender's company
1904 oversees enlargement of
Greenland Docks

The Prince of Wales (1841–1910) opened London's Tower Bridge on June 30, 1894. Enthusiasm for the new bridge was muted, however, and the publishers of the trade journal *The Builder* even refused to print an image, not wanting to waste plates on it. Yet the bridge was a dream come true for the million-odd Londoners who lived south and east of the Thames as it meant an end to the daily nightmare of using the hopelessly congested London Bridge.

To mark the 50th anniversary of Queen Victoria's accession to the throne (she reigned from 1837–1901), Prince Edward laid the foundation stone on June 21, 1886. The bridge had been designed by the late Sir Horace Jones, but the actual construction was led by Sir John Wolfe-Barry. While Wolfe-Barry had enjoyed greater artistic freedom following Jones' death, he nevertheless followed the suggestion of cladding the steel structure in Portland stone and granite from Cornwall. These mate-rials ensured that the bridge's two towers responded sty-listically to the nearby medieval Tower of London and the neo-Gothic Houses of Parliament. The borrowings from medieval fortified architecture would later prompt H.G. Wells (1866–1946) to describe Tower Bridge as a "Stock-

broker in Armor." What really made the bridge a wonder of Victorian engineering was its hydraulic machinery, designed by Armstrong Mitchell & Co., housed in the base of the towers. It could raise both bascules, weigh-ing 907 tons, within 90 seconds—and did so 6,160 times in 1895 alone. Large sailing ships were thus still able to pass beneath the high-level footbridges into the Pool of London.

The bridge made headline news in 1912 when Frank McClean flew his aircraft between its towers. Forty years later, bus driver Albert Gunter and his passengers had a lucky escape when the bridge started to open with the bus still on it; the driver put his foot down and got his bus on to the other side in the nick of time. Even the architecture critic Eric de Maré, who had once spoken disparagingly of the Tower Bridge, eventually had to admit that it did its duty with admirable regularity and efficiency. It is little surprise, then, that in 1970 the city fathers of Lake Havasu City in Arizona wanted to buy this great symbol of British ingenuity. They made a mistake, however, and instead bought London Bridge (*c.* 1820) that was dismantled and re-built stone by stone in the U.S.A.

The hydraulic machinery within Tower Bridge
was electrified in 1976

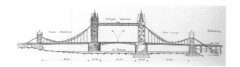

Tower Bridge has a span of 270 meters (design drawing at top).
It was once described as a "great symbol of British ingenuity"

A triumph of stone and steel
Dresden: The "Blue Wonder" 1891–93

"The breast swells proudly/The gaze rests proudly on unveiled glory./Future generations of Saxons will know what quarries and mines/In their land contain./It will in future extol the art of their fathers/A triumph of stone and steel!"

A. von Wendell's hymn for the opening of the "Blue Wonder"
July 15, 1893

Claus Köpcke
(1831–1911)

1869 Professor of Design and Construction of Railways, Roads and Hydraulic Engineering at Dresden Polytechnic
1872 Consultant to the Royal Saxon Ministry of Finance
1903 retires from the Saxon civil service having achieved grade of assistant secretary
1911 dies November 21

In 1893, von Wendell expressed the great pride of the region at the official opening of the King Albert Bridge over the Elbe river, for now that a bridge was available to travellers, the ferry between Loschwitz and Blasewitz could be bypassed and the court city of Dresden could be reached far more quickly. Nearly twenty years earlier, two designs had been submitted for the new Elbe crossing. It was another ten years after that before the Royal Saxon Ministry of Finance, prompted by a petition of Loschwitz residents, awarded the "Sächsische Eisenbahnkompanie AG" (Saxon Railway Company) the privilege to build the bridge.

The Saxon Railway Company, a local steel company, had adopted Professor Claus Köpcke's design for a reinforced suspension bridge with three spring-loaded joints in the main span. The "Sächsische Eisenbahnkompanie AG" began production on April 1, 1891 under the supervision of building inspector Hans Manfred Krüger. Ninety-seven tons of rivets and 3,800 tons of steel were transported by rail and ship from the foundry in Cainsdorf to the construction site. Krüger was also in charge of the erection of the steel framework, which took place over the next nine months. The approaches to the bridge and

its anchorages, on the other hand, were the responsibility of Aemil Hugo Ringel. On July 11, 1893, the bridge successfully underwent a loading test that cleared the way for an official opening four days later.

Although the "King Albert Bridge" was at first reviled as an eyesore, it quickly came to be regarded as a technological masterpiece whose colour soon had locals calling it the "Blue Wonder." Willy von Weyern recorded new public sentiment just a few years later: "At Blasewitz there is a bridge/Whose colour blue once pleased the eye./ 'A Blue Wonder,' rose the enthusiastic cry./ And wondered soon where the bridge's blue had gone."

Exposed to the elements for a few years, its blue color had turned into an unattractive shade of grey that was then painted over with yellow. Only in 1952 was the original color of the "Blue Wonder" restored.

Tested with an evenly distributed load of 157 tons, the "Blue Wonder" moved only 9 mm on July 11, 1893

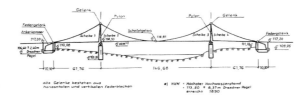

left: Design drawing of the Loschwitz Bridge

The Loschwitz Bridge across the Elbe has a span of 141.5 m and is 260 m long; it is known far beyond Saxony as the "Blue Wonder"

To connect nations and ages
Paris: Alexander III Bridge 1896–1900

"This bridge will form an endless arch from the century ending to the one beginning and is built to connect nations and ages...France dedicates it to your father, Alexander. Complete his work, inheritor of his fame. Take the means into your loyal hand...and with this steel hammer work gold and ivory."

Declaimed by Paul Mounet as the foundation stone
of the Alexander III Bridge was laid, 1896

Louis-Jean Résal
(1854–1920)

1854 born in Besançon trains as a bridge engineer
1893 constructs the three-arch steel bridge at Mirabeau; its side spans measure 32 meters, its main span measures 93 meters; A. Injalbert designs four sea gods for it
1897–1900 construction of the Pont Alexandre III in collaboration with the engineer d'Alby and sculptors E. Frémiet, G. Michel, A. Lenoir, P. Granet, C. Steiner, J. Coutan, L. Marqueste, J. Dalou and H. Gauquic
1900 the 120-meter-long Debilly Bridge is completed by engineers d'Alby and Lion and builders Dayde and Pille on April 13, 1900

On October 7, 1896 Paul Mounet, the renowned actor at the Comédie Française, directed his solemn words to tsar Nicholas II of Russia (reigned 1894–1917) as he laid the foundation stone for the Alexander III Bridge. The name of the bridge was to be a permanent reminder of the late tsar Alexander III (reigned 1881–94) and of French-Russian friendship. In 1892, the tsar had formed an alliance with the Third Republic under Marie François Sadi Carnot (in office 1887–94) against the German emperor William II (reigned 1888–1918).

The Third Republic's political alignment, technological capabilities and economic strength were to be expressed through this ingenious Seine bridge. It fell to the engineer Louis-Jean Résal (1854–1920), who had already built the Mirabeau steel bridge in the city in 1893, to realize this engineering feat in collaboration with his colleague d'Alby. Résal's very shallow arches spanning 107.5 meters did indeed demonstrate the skill of French bridge engineers. He deliberately positioned the floral ornamentation and garlands below the balustrade so that the deck's fifteen three-pinned steel arches remained visible. This required particularly strong abutments, whose bases at either side were provided by four

18-meter-high piers that allowed representative statuary by several sculptors to be realized. The bases also provided space for depictions of glorious eras in French history, for instance that of Charlemagne and the Renaissance. The less glorious age associated with the French Revolution was overlooked. Copies of candelabra from St. Petersburg led to the center of the bridge where the French and Russian coats of arms were flanked by personifications of the Neva and the Seine. During the Exposition Universelle, the Alexander III Bridge proved highly effective as a triumphal bridge and national monument. Criticism that its ornamentation was too irregular was short-lived. Countless experts from near and far admired the pioneering achievement of its steel construction as they strolled from the Place de Clemenceau on the Champs Elysées past the newly-opened Grand Palais and Petit Palais to the Hôtel des Invalides.

Tsar Nicholas II attending the foundation ceremony for the Alexander III Bridge, Musée de la Batellerie, Paris, late nineteenth century

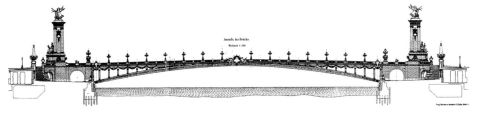

Schematic drawing of the Pont Alexandre III

The Alexander III Bridge was built for the 1900 Exposition Universelle to the glory of war and peace, to commemorate the alliance between Russia and France and as the epitome of technology in the Third Republic, 1896–1900

Best and cheapest
Quebec City: Quebec Bridge 1904–07, 1913–17

"I, Theodore Cooper, hereby conclude and report, that the cantilever superstructure plan of the Phoenix Bridge Company is the best and cheapest plan and proposal."

Theodore Cooper in his report to the
Quebec Bridge Building Company, June 23, 1899

Theodore Cooper
(1839–1919)

1858 earns an engineering degree
from Rensselaer Polytechnic
1861–72 drafted into the U.S. Navy
as an engineering assistant on
board the gunboat Chocura in the
last three years of the American
Civil War
1872–74 serves as chief engineer
under bridge engineer James
Buchanan Eads (1820–87) during
construction of the Mississippi
Bridge at St. Louis
1879 superintendent of the
Andrew Carnegie Keystone Bridge
Building Co., Pittsburgh
designs the bridges at Seekonk
in Providence, Sixth Street in
Pittsburgh and Second Avenue
in New York
1907 collapse of the Quebec
Bridge ends his career
1919 dies in New York on
August 24

At first, the discussions about the construction of the Quebec Bridge across the St. Lawrence River were no different from other projects of the day. In 1897, the Quebec Bridge Building Co. approached Theodore Cooper (1839–1919) and asked him to give his expert opinion on the designs of the Pennsylvania-based Phoenix Bridge Building Company. He decided in favor of the design by Peter Szlapka, Phoenix's chief designer, because it was the best and the cheapest option. The Quebec Company did not doubt Cooper's integrity and appointed him chief engineer on May 6, 1900.

Five days earlier, Cooper had lengthened the main span from 488 to 549 meters without taking increased wind loading into account. When his figures were checked, he replied, "There is no one competent enough to criticize us." Cooper was on site on only three occasions. When S.N. Parent, the head of the Quebec Bridge Building Co., asked him in 1904 when he would visit the site, Cooper made his apologies, claiming to be stressed and ill. A young engineer named Norman McClure deputized for Cooper on site after that. In reply to McClure's concerns expressed in letters to Cooper, the latter replied: "Make as good work of it as you can, it is not serious."

In August 1907, when the southern cantilever arm was already 230 meters long, buckling was discovered in the main beams near the base of the rectangular towers. Cooper's instruction to halt work reached John Sterling Deans, the chief engineer of the Phoenix Bridge Building Company, on August 29 at 3 pm. He failed to act on it. Two hours later, the bridge collapsed in less than 15 seconds, taking 85 men with it. Only 11 of them survived.

Too long a span, excessive cost-saving, insufficient materials testing, inappropriate on-site supervision and the absence of the elderly bridge designer were the factors that led to the disaster. Although a Royal Commission found John Deans and the Quebec Bridge Building Company culpable, blame was heaped on Cooper. His eminent career came to an abrupt end and he was forced to retire from public life. On September 12, 1916 an attempt was made to position a new central suspended span weighing 5,200 tons between the cantilever arms —but in vain. One of the jacks failed, causing the whole structure to crash into the river, claiming eleven more lives. The Prince of Wales finally opened the Quebec Bridge to traffic in August 1919, the month Theodore Cooper died.

September 12, 1916, collapse of the center span of
the reinforced Quebec Bridge

The Quebec Bridge, the longest cantilever bridge in the world,
both famous and notorious for failing in 1907 and 1916

The Old Coat Hanger
Sydney Harbour Bridge 1926–32

"The granite-faced abutment towers and pylons, simple and elegant, are the architectural features of the Harbour Bridge which would otherwise be an immense utilitarian steel structure."

John Job Carew Bradfield on the Harbour Bridge, pre-1932

John Job Carew Bradfield (1867–1943)

1867 born December 26 in Sandgate by Brisbane
1890 technical drawer employed in The Office of Public Works, New South Wales
1895 designs Burrunjuck and Cataract dam in New South Wales
1912/13 engineer-in-chief on the Harbour Bridge and the Metropolitan Railway, Sydney
1934–40 construction of the Story Bridge near Brisbane
1943 dies September 23, and is buried in St. Johns

Sir Ralph Freeman (1880–1950)

1880 born November 27
1905 African Bridge across the Zambezi at Victoria Falls, Zimbabwe
1924–32 Sydney Harbour Bridge
1933 design for the Kobri Kasr el Nil Bridge, Egypt
construction of the Tyne Bridge, Newcastle
construction of five bridges in Rhodesia (now Zimbabwe)
ships built in Lancashire
1950 dies March 11

For civil engineer John Job Carew Bradfield (1867–1943), it was never enough that structural considerations alone should determine the appearance of a bridge. He developed this belief while working on the design of Sydney Harbour Bridge and continued to do so on the Story Bridge in Brisbane, but the origins of Sydney Harbour Bridge date back before Bradfield was able to publicize his ideas and realize them.

In 1911, New South Wales premier William Holman announced that his Cabinet had chosen a design for a bridge that would carry both tram and vehicular traffic as well as pedestrians. Despite efforts to get the project underway, negotiations dragged on. In 1922, the New South Wales government announced another worldwide competition. Six different companies submitted twenty designs, including one from Bradfield himself. Taking his inspiration from Gustav Lindenthal's 1916 Hell Gate Bridge in New York, Bradfield rejected the idea of a cantilever bridge in favor of an arch construction. The final decision was taken on March 24, 1924: the winning submission was that of the English firm Dorman Long and Co. in Middlesborough, whose design was for a two-hinged arch. Ralph Freeman (1880–1950) in London

was contracted to make several design improvements, adding another two years before work on the foundations started. In November 1929, the two arch-halves were cantilevered out from either side until they joined in the middle, on August 20, 1930. Even then the characteristic arch profile was unmistakable and would later earn Sydney Harbour Bridge the nickname of the "Old Coat Hanger."

The massive deck, measuring 49 meters across, was next constructed as well as the two towers that are reminiscent of Egyptian architecture. Work on the bridge was finally completed in February 1932. Ninety-six steam trains tested the strength of Sydney Harbour Bridge at various positions before New South Wales premier John Thomas Lang (in office 1925–27 and 1930–32) opened it to traffic on March 19, 1932. Through bridge tolls, it took until 1988 to pay off the cost of construction (A\$13. 5 million), twelve years after the billionth vehicle had crossed the bridge. In addition to construction costs, a considerable amount also had to be spent on maintenance. Among the engineers working on the bridge was Paul Hogan, a man who later turned his hand to acting and who found fame as "Crocodile Dundee."

View of the steel struts

New York's Hell Gate Bridge was the model for Sydney's Harbour Bridge

At 1,149 meters long and with a main span of 503 meters, Sydney Harbour Bridge was the world's longest steel arch bridge when it was completed

Noah's Ark Returned
New York: George Washington Bridge 1927–31

"The George Washington Bridge over the Hudson River is the most beautiful bridge in the world. Made of cables and steel beams, it gleams in the sky like Noah's Ark returned. It is blessed. It is the only place of grace in this disordered city…When your car moves up the ramp, the two towers rise so high that they bring you happiness; their structure is so pure, so regular that here, finally, steel architecture seems to laugh."

Le Corbusier, *When the Cathedrals Were White*, 1937

**Othmar Hermann Ammann
(1879–1965)**

1879 born March 26 near the Swiss town of Schaffhausen, son of a hatmaker and businessman
1897–1903 studies civil engineering at Zurich Polytechnic works on the construction of 30 steel railway bridges
1908 investigates the causes of the collapse of the Quebec Bridge
1912–23 chief assistant to Gustav Lindenthal, designer of Hell Gate Bridge across the East River
1931–33 construction of Bayonne Bridge and Triborough Bridge
1937 completion of the Lincoln Tunnel
1939–64 completion of the Outerbridge, Goethals, Queensboro, Bronx Whitestone and Throgs Neck bridges
1965 dies at his home in Rye, New York on September 22

There was great delight on October 24, 1931 when Franklin D. Roosevelt opened the bridge across the Hudson River, naming it after the first American president and praising it as "almost superhuman in its perfection." Its designer, Othmar Hermann Ammann (1879–1965) had, for the first time in history, succeeded in spanning a distance of more than 1,000 meters, thereby creating what was then the world's longest suspension bridge.

As evidenced by architect, painter, and writer Le Corbusier's (1887–1965) praise, many found the George Washington Bridge attractive. No one was particularly concerned that Ammann had planned to give the bridge a completely different appearance. The steel skeletons of both 184-meter-high towers were originally to have been cast in concrete and clad in granite, but following the Wall Street Crash of 1929, the Port of New York Authority no longer had sufficient funds at its disposal. Ammann saved an additional 3.5 million dollars by rejecting chain suspension, favouring instead a cable suspension system. Like Roebling on the Brooklyn Bridge before him, he made use of cable-spinning technology. The lower of the bridge's two decks was originally to have included railway lines for suburban trains, but the

decision was taken to abandon them, even though the eight lanes for vehicle and two pedestrian walkways on the upper deck had already been completed. New York's city fathers were grateful for its presence when they decided to add a further six traffic lanes on the lower deck between 1958 and 1962. The volume of traffic in the city had expanded enormously due to the success of Henry Ford's idea of the mass manufacture of cars. Ford sold fifteen million of the model nicknamed "Tin Lizzie." In 1932, 5.5 million vehicles used the George Washington Bridge; twenty years later, the figure had risen to 28 million.

Ammann was criticized in Europe for eschewing the traditional stabilizing truss on most suspension bridges. He was convinced the sheer weight of the suspension bridge ensured sufficient stability. He was proved right in December 1965 when a private aircraft crashed into the bridge, leaving both it and the pilot unharmed. Ammann's European critics were finally silenced, after going so far as to say, "Ammann in America is a fool if he builds a suspension bridge that size with no stiffening truss."

View of the carriageway on the George Washington Bridge

The towers of the George Washington Bridge were originally going to be covered, but the stock market crash made financing scarce

In all its majesty
San Francisco: Golden Gate Bridge 1933–37

"Strauss will never build his bridge, no one can bridge the Golden Gate because of insurmountable difficulties which are apparent to all who give thought to the idea."

John Bernard McGloin, S.J., "Symphonies in Steel: San Francisco Bay Bridge and the Golden Gate," *San Francisco, the Story of a City*, 1978.

Joseph Baermann Strauss
(1870–1938)

1870 born January 7 in Cincinnati, Ohio
1929–37 engineer-in-chief on the Golden Gate Bridge
1938 dies May 16 in Los Angeles

Irving Foster Morrow
(1884–1952)

1884 born in Oakland, California
1900 studies architecture at the University of California at Berkeley architecture studies at the École des Beaux-Arts in Paris
1918 opens an architect's practice with his wife Gertrude Comfort Morrow, Berkeley's first female architecture graduate
1930–37 designs Golden Gate Bridge
1952 dies October 28 on a bus journey through San Francisco

It must have given engineer-in-chief Joseph B. Strauss (1870–1938) great satisfaction to answer his critics in the May 27, 1937 edition of the *San Francisco Chronicle* when he said the Golden Gate stood "in all its majesty" and to have the President Franklin D. Roosevelt, speaking from the White House, inform the world of the opening of the Golden Gate Bridge. Strauss had endured a long and complicated road to build the bridge linking San Francisco and Marin County.

A bridge over the bay was a long time in coming. In 1844, the mayor of Yerba Buena, the first Spanish settlement on what became San Francisco, demanded a bridge. In 1872, discussion was re-ignited by the railway tycoon Charles Crocker. Yet nothing was accomplished for nearly forty years, when Joseph B. Strauss' design was vetoed because of ferry operators' concerns at possible financial losses coupled with doubts about the design's structural integrity and aesthetics. The establishment of the partly government-run Golden Gate Bridge Authority in 1928 was intended to drive the development of the bridge, however, and Professor Charles Alton Ellis was called in to validate Strauss' calculations. Technical advisor and architect Irving Foster Morrow

(1884–1952) added to the design's attractiveness by giving the huge towers a sense of lightness by designing them in the Art Deco style and, as a contrast to the "delicate" steel structure, designed impressive concrete anchorages for the bridge cables. Perhaps most famously, Morrow gave the bridge its color, International Orange, which Strauss described as reminiscent of the Grand Canyon.

Strauss was able to proceed and he took most of the credit for the Golden Gate Bridge, only indirectly acknowledging Morrow's achievement. Nor did he ever mention the men who, after 1933, solved the structural problems caused by the tides and storms, such as those encountered with a cofferdam on the northern Marin Tower and a massive fender at the south tower that stood in 34 meters of water. Strauss' achievements did include the use of a safety net that saved the lives of 19 men. Since it opened on May 27, 1937, more than 4.1 billion vehicles have crossed the 2,727-meter Golden Gate Bridge that "looms mountain-high" as Strauss wrote in his ode titled *The Mighty Task Is Done*.

The Golden Gate Bridge is famous for its color, International Orange

Across the Valley of Death
The Kwai Bridge 1943–45

"...putting all their reduced weight behind it, contributing the additional sacrifice of this painful effort to the sum total of suffering which was slowly bringing the River Kwai bridge to a successful conclusion."

Pierre Boulle, *Le Pont de la Rivière Kwaï*, 1952

Pierre Boulle
(1912–1994)

1912 born in Avignon
1939–40 soldier with the French army in Indo-China during World War II
1940–44 soldier with the Free French in Singapore, captured and tortured by gendarmes of the Vichy regime
1952 *The Bridge on the River Kwai, Planet of the Apes*
1978 *The Good Leviathan*
1994 dies aged 81

David Lean
(1908–1991)

1908 born March 25 in South Croydon
1948 *Oliver Twist*
1957 *The Bridge on the River Kwai*
1962 *Lawrence of Arabia*
1965 *Dr. Zhivago*
1970 *Ryan's Daughter*
1984 *A Passage to India*

The French author Pierre Boulle (1912–94) could not have given a more poignant description of the fate of British soldiers in his 1952 novel *Le Pont de la Rivière Kwaï* (The Bridge on the River Kwai). It describes how, during World War II, the Japanese forced British and other Allied prisoners of war to drive a strategically important railway line through the Thai jungle and to build a bridge across the river Kwai. A British demolition squad, made up of idealists and fanatics, wanted at all costs to prevent the bridge's completion. The British officer Colonel Nicholson, whose pride the merciless Japanese Colonel Saito couldn't break, not only took charge of the construction of the bridge, but also came to regard it as a personal challenge to see the daring structure completed. He felt that nobody had the right to destroy the monument, not even his own men. A bitter struggle between life and death began in the rainforest. Patriotism and blindness, loyalty and mistaken aims collided. Boulle chose to end his novel with the words: "The Japs fired back. Soon the smoke spread and crept up as far as us, more or less blotting out the valley and the River Kwai. We were cut off in a stinking grey fog." In 1956/57, his novel was filmed by Englishman David Lean (1908–91), whose mas-

terly direction and Alec Guinness' magnificent performance as Colonel Nicholson made the film a box-office hit around the world. Based on wartime events, the film won seven Academy Awards.

Work on the real railway bridge across the river Kwai began in October 1942 in the Thai province of Kanchanaburi northwest of Bangkok. The steel construction had been dismantled on Java by the Japanese Army, shipped to Thailand and was re-built using the forced labor of thousands of British and Allied POWs and locals. It was part of a 425-km-long railway line carrying Japanese supplies from Ban Pong in Thailand to Thanyuzayat in Burma. It was an outstand-ing engineering achievement that was completed within 16 months.

A Dutch POW wrote: "Hardly a day went by when we didn't have to bury one of our men. The Japanese increased the workload every day and beatings were a daily occurrence." In all, 18,000 Allied POWs and 78,000 Asian forced laborers died needlessly. The bridge they built across the valley of death stood only for a few months: on October 13, 1945, the U.S. Air Force largely destroyed it, only its semicircular arches remaining intact.

Alec Guinness as Colonel Nicholson and Sessue Hayakawa as Colonel Saito before the Hollywood version of the bridge, in David Lean's *The Bridge on the River Kwai*, 1957

The Kwai Bridge was destroyed in 1945 and re-built; it later came to worldwide attention through David Lean's film

For Peace and Unity
Innsbruck: The "Europabrücke" 1959–63

"For millennia, the Brenner has been one of the main trans-Alpine routes connecting the north and the south. Where in prehistoric times there were only bridle paths, now there is a motorway. With the free nations of Europe uniting ever more closely, let the Brenner autobahn and the 'Europabrücke' be not only a means allowing the movement of traffic; may they also be the symbol of a peaceful future."

Inscription on the commemorative plaque on the
stairway leading to the Europa Chapel, 1963.

Karl Plattner
(1919–1986)

1919 born in Mals, South Tyrol
1947–50 studies in Milan, Florence and Paris
1950 municipal prize awarded by the Town of Rovigo, ex aequo
1950/51 series of frescoes in Mals and Naturns
1954/55 series of frescoes for the seat of the provincial government in Bolzano
1959 wins Industrial Union Prize, Bergamo
1963/4 frescoes in the Europa Chapel
1964 wins first prize for his print *Premio Suzzara*
1986 dies in Milan

Karl Plattner in front of his fresco, *Rape of Europe*, 1964

The bells of the Europa Chapel first rang out across the region on November 17, 1963 to announce the opening of the "Europabrücke." Its inscriptions "North and South Tyrol welcome Europe" and "For Peace and Unity" attest to the enthusiasm that was felt at the time. No one would again have to experience the strain of journeys that carters faced in earlier centuries as their horse-drawn carriages trundled over the jarring cobblestones of the roads built by the Romans.

Construction work on the "Europabrücke" began in April 1959, with the first sod being turned by Federal Minister Fritz Bode and provincial governor Hans Tschigg-frey during the Chancellorship of Julius Raab (1953–61). In only four years, the Wipp Valley was spanned by Europe's tallest pier bridge with a stiffened steel box girder and a 22.8-meter-wide steel plate deck.

Twenty-three men lost their lives building the bridge, including the bridge engineers Herbert Henning, Hermann Staf and Hans Gaidoschik. In 1960, therefore, the provincial government of Tyrol made plans to build a commemorative Europa Chapel next to the bridge and commissioned a design from a local architect called Hubert Prachensky. He built a V-shaped room ending in a window with a view of the bridge where the men died. The South Tyrolean artist Karl Plattner (1919–86) designed a set of frescoes for the space: starting on the left-hand side of the Chapel, he showed the workmen and St. Christopher carrying the young Christ across a river—much the same way the "Europabrücke" carries travellers. Next follows a woman holding one of the victims in her arms. The series of frescoes on this side ends with the Riders of the Apocalypse storming across the bridge. Only Saint John of Nepomuk, patron saint of bridges, can stop them. The frescoes on the right-hand side start with scenes from Tyrolean history: a shepherd boy looks on as the Peasants' War is waged under the banners of Austria and Tyrol. The final image is taken from Greek mythology and shows Zeus in the form of a bull carrying off Europa, thus making a connection with the name of the "Europabrücke," the "Gateway to the North and the South," Europe's artery serving long-distance trade and international tourism.

Hubert Prachensky, sketch of the Chapel beside the "Europabrücke," 1963

This sketch shows how the bridge merges into the landscape

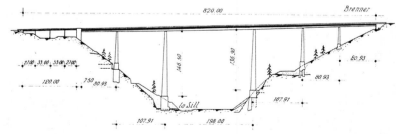

The "Europabrücke," Europe's highest pier bridge: 820 m long, 190 m high and with a maximum span of 198 m.

In harmony with the magnificent landscape

New York: Verrazano Narrows Bridge 1959–64

"The width and depth of the double deck, the sag of the cables, and the height of the towers, resulted in a well-balanced design of light and graceful appearance, in harmony with the magnificent landscape. The bridge will form a monumental portal at the entrance of the New York Harbor."

Othmar Hermann Ammann, 1963

Othmar Hermann Ammann (1879–1965)

1879 born March 26 near the Swiss town of Schaffhausen, son of a hat-maker and businessman
1897–1903 studies civil engineering at Zurich Polytechnic works on the construction of 30 steel railway bridges
1908 investigates the causes of the collapse of the Quebec Bridge
1912–23 chief assistant to Gustav Lindenthal, designer of Hell Gate Bridge across the East River
1931–33 construction of Bayonne Bridge and Triborough Bridge
1937 completion of the Lincoln Tunnel
1939–64 completion of the Outerbridge, Goethals, Queensboro, Bronx Whitestone and Throgs Neck bridges
1965 dies at his home in Rye, New York on September 22

The Italian navigator Giovanni da Verrazano (1485–1528) could not in his wildest dreams have imagined that it would one day be possible to build a bridge across the North American bay that he discovered on April 17, 1524 and, even less so, that it would bear his name. From his childhood, he knew bridges only the size of Florence's Ponte Vecchio. Sailing in the Dauphine under the flag of king Francis I of France (reigned 1515–47), Verrazano was the first European to land in the bay where, centuries later, Brooklyn and Richmond (Staten Island) would become part of New York City.

What was unthinkable to Verrazano became reality thanks to no less a figure than the internationally acclaimed designer of the George Washington Bridge, Othmar Hermann Ammann (1879–1965). More than any-one else, he opened up New York Bay to traffic: it was he who designed and built a host of bridges there— Throgs Neck, Bronx Whitestone, Triborough, Queens-boro, Bayonne, Goethals and the Outerbridge Crossing as well as the Lincoln Tunnel. Aged 75, he was now set to build a fixed link between Brooklyn and Staten Island, a project that was to mark the pinnacle of his illustrious career. His firm of civil engineers, Ammann & Whitney,

was commissioned to build it by the "Triborough Bridge and Tunnel Authority." Up to 1,200 construction workers were employed on the job at one time. The two cellular 211-meter-high towers each weighed 27,000 tons; even if they are a repetition of the Bronx Whitestone Bridge, and their cross beams appear bulky and top-heavy compared with the delicate structure as a whole, on its completion the Verrazano Narrows Bridge broke all the records. It had the world's longest span and carried twelve carriageways on two decks. It was also the heaviest bridge, using more steel than any other bridge on earth.

On November 13, 1964, the year the Verrazano Narrows Bridge opened, president Lyndon B. Johnson (in office 1963–69) awarded Ammann the "National Medal of Science" for his life's work. The Swiss-born bridge engineer died a year later. Employees of Ammann & Whitney had a commemorative plaque inscribed in memory of their boss; it reads: "Othmar H. Ammann, a man endowed with dignity, humility and integrity, an engineer gifted with rare insight and leadership, this plaque is presented in fond memory by his employees."

Marathon runners cross the Verrazano Narrows Bridge

Spinning the suspension cable on the Verrazano Narrows Bridge

The world's longest single span on its completion, the Verrazano Narrows Bridge measures 1,298.45 meters

Connecting what the waters once divided
Panama: Bridge of the Americas/Puente de las Américas   1962

"The bridge over the Panama Canal shall bear the name Bridge of the Americas. Said name will be used exclusively to identify said bridge. Panamanian government officials shall reject any document in which reference is made to the bridge by name other than Bridge of the Americas."

National Assembly Resolution of October 2, 1962

Panama Canal

Spanish settlers first conceived of cutting through the Panama isthmus towards the late sixteenth century
1846   The United States secures territorial sovereignty in an agreement with New Granada (Colombia since 1862)
1879   The Compagnie Universelle du Canal Interocéanique under Ferdinand de Lesseps, the engineer of the Suez Canal, begins construction of a sea-level canal
1889   The project fails for technical reasons
1901/03   The United States and the Republic of Panama sign agreements on territorial sovereignty and rights of passage
1914   Panama Canal opens on August 15
1977–79   Decision taken to introduce joint U.S. and Panamanian administration of the Canal and the Canal Zone (16km-wide strip of land on either side of the Canal) until 1999
1999   Panama assumes sole control of the canal

The Bridge of the Americas connects North and South America in more ways than one. Near the Pacific port terminal of Balboa, and here alone, a bridge crosses the Panama Canal. The canal was completed by the United States on August 15, 1914 and, although it was described as the Eighth Wonder of the World, it caused a man-made division of North and South America. When completed in 1962, the Bridge of the Americas somewhat abated the effects of that separation.

A late reminder of the pan-American movement, the bridge became part of the Pan-American Highway, the main road network linking North and South America. It replaced an old ferry known as the "Thatcher Ferry" at the same point. To the annoyance of Panama, the Americans liked to refer to it rather disparagingly as the "Thatcher Ferry Bridge." The Panamanian National Assembly had an answer for that.

Plans to build a bridge across the canal were first made in 1923. Further inland, at the two sets of locks at Miraflores, a small swing bridge was built in 1942, but it took until 1955 before President Eisenhower signed the agreement to build the Bridge of the Americas. The contract, worth more than 20 million dollars, was won by

John F. Beasly & Company. The 1,655-meter-long steel arch bridge opened on schedule on October 12, 1962. With clearance of 106 meters, even the largest freighters can pass beneath it. The Bridge of the Americas was unable to cope with the increase in traffic using the Pan-American Highway and, in 1998, an extra lane was added to it, slightly easing congestion. Since then, plans have been under discussion to build a second bridge or a tunnel located further west. For the time being, the Bridge of the Americas remains the western hemispheres' only inter-continental bridge.

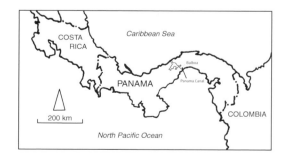

Ships sailing between the east and west coasts of the United States have had their journeys cut by 8,000 nautical miles since the Panama Canal opened in 1914. The Canal is 81.6 km long and is between 90 and 300 m wide (journey time 7–8 hours). Between 1962–71, its locks were enlarged and its minimum depths and widths were increased

The deep arch of the Bridge of the Americas is reminiscent of the arches of ancient stone bridges

Symbol of the Revolution of the Carnations
Lisbon: Ponte de 25 Abril 1962–66

"The Portuguese Army calls on the residents of Lisbon to stay at home and to remain calm."

Announcement on Portugal's "Clube"
radio station, April 25, 1974

David Barnard Steinman
(1886–1960)

from 1886 trains at City College and
Columbia University, New York City
1916 assists Gustav Lindenthal
on Hell Gate Bridge, New York
1926 constructs the Florianópolis
Bridge in Brazil together with his
partner Holton D. Robinson
400 other bridge designs follow,
including:
1927 Carquinez Strait Bridge,
California
1929 Mount Hope Bridge, Rhode
Island
1931 St. John's Bridge, Oregon
1936 Henry Hudson Bridge,
New York
1952–56 Kingston Bridge
1953–57 Mackinac Bridge between
Sault Ste. Marie and Michigan
1960 advises on the design of the
Tagus Bridge near Lisbon

On April 25, 1974, the "Salazar Bridge" across the river Tejo had been secured by the military, whose aim was to free Portugal from decades of dictatorship. The bridge's original name, "Salazar Bridge," was discouraged after the coup as it was a reminder of the dictator António de Oliveira Salazar, who acted as Portugal's prime minister from 1932–68. The new government re-named the bridge the "Ponte de 25 Abril" in honor of the country's Liberation Day.

The bridge's history remains closely associated with the Portuguese dictatorship, however. In 1953, Salazar set up a project office known as the "Gabinete da Ponte sobre o Tejo." In the late 1950s, the project office and its director, engineer Jose do Canto Moniz, announced an international competition. Only the Pittsburgh-based United States Steel Corporation was felt to be sufficiently competent to undertake the project, due to its partnership with "Steinman Boynton Gronquist." The advice of the renowned bridge designer David Barnard Steinman (1886–1962) was thus secured for the project. His scientific studies into aerodynamic stability permitted the construction of lengthy suspension bridges and were favorably received in Lisbon. Moniz could not know that

Steinman would die later that year and that his partner Ray M. Boynton would have to take on the design duties.

Construction work started in 1962. Despite the similarities with the Golden Gate Bridge, new technologies were employed. For the first time in Europe, compressed-air domes were used to control the stability and sinking of the caissons. This enabled engineers to found one of the piers on bedrock 79 meters below water level, a depth never previously reached. To ensure stability, the world's longest continuous truss measuring 2,277 meters was built. Another unusual feature was the deep truss of transverse X-stays that could later be adapted to allow the construction of a railway. Suspended from the 190.5-meter-high towers, the truss allowed seventy meters of clearance to ensure that even the largest ships had access to the port of Lisbon.

When the lower deck finally opened to railway traffic in 2000, the Tejo Bridge had long come to symbolize Portugal's 1974 Revolution of the Carnations.

Citizens celebrating the Revolution of the Carnations on April 25, 1974

With a main span of 1,013 meters and side spans of 483 meters, the Tejo Bridge (Ponte de 25 Abril)
was mainland Europe's longest suspension bridge upon its completion

The Gateway to the East
Istanbul: Bosporus Bridge (Bogaziçi) 1970–73

"It has come to my attention that you intended to construct a bridge from Pera to Constantinople, but that it was not built for lack of an expert. I am able to design a bridge such that allows a ship in full sail to pass beneath it."

Leonardo da Vinci in a letter to Sultan Bejazid II Wali, 1502

Ever since antiquity, men's minds have been exercised by the idea of bridging the Bosporus and the Dardanelles, the straits connecting the Black Sea and the Sea of Marmara with the Aegean. The dream was to create a permanent link between the Occident and the Orient. The pontoon bridge built across the Bosporus in 513 B.C. by the Persian master builder Mandrokles of Samos for the Scythian campaign of the Persian king Darius I (reigned 522–486 B.C.) is remembered to this day. Thirty-three years later, Egyptian and Phoenician engineers made attempts to bridge the Hellespont, but according to the Greek historian Herodotus (c. 485–425 B.C.) their pontoon bridges were destroyed in a storm. In his anger, Xerxes I (reigned 486–465 B.C.) had the engineers beheaded and ordered new pontoon bridges to be built to enable him finally to attack his sworn enemies, the Greeks. When Sultan Mehmed II Fatih (reigned 1451–81) conquered Constantinople almost 2,000 years later and made it the capital of the Ottoman Empire on May 29, 1453, fixed bridges were still not in place.

For Mehmed's successor, Sultan Bejazid II Wali (reigned 1481–1512), polymath Leonardo da Vinci (1452–1519) designed a 360-meter-long bridge in 1502

that was to connect Constantinople with Pera, now the suburbs of Eminönü and Beyolu. The Sultan believed da Vinci's highly original design to be unworkable and abandoned the project. The design was eventually used—in 2001, when Norwegian architects Selberg Arkitektkontor AS realized it as a wooden pedestrian bridge, albeit on a reduced scale of 57 meters, in Ås near Oslo. Sultan Bejazid was mistaken: computer simulations have now shown that Leonardo's design for a bridge over the Golden Horn was structurally sound.

Centuries would pass before the Bosporus was bridged. Between 1970 and 1973, the first fixed link between Europe and Asia was created in the form of the Bosporus suspension bridge, built by the German engineering company Hochtief AG under its chairman, Enno Vocke, and engineer Celalettin Dursun in collaboration with the British firm of Freeman, Fox and Partners. Proceeds from bridge tolls were so lucrative that in 1988 another bridge, the Fatih-Sultan-Mehmet Bridge (Bosporus II), with a span of 1,090 meters, was built. A privately financed third gateway to the Orient is now planned.

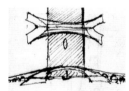

Leonardo da Vinci, *Self-Portrait*, Biblioteca Reale, Turin, around 1512

Design by Leonardo da Vinci, Ms. L, fol. 66r, Bibliothèque de l'Institut de France, Paris, 1498–1502

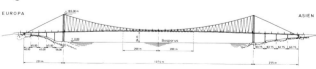

Design drawing of the Bosporus Bridge, 1970–73

The first bridge linking the Occident and the Orient: the Bosporus I Bridge has a span of 1,074 meters and clearance of 64 meters, 1970–73

A Triumphal Arch in the Port of Hamburg

Hamburg: The Köhlbrand Bridge 1970–74

*"The world has many wonders, so the books say/From the Pyramids of Egypt to the Gardens of Babylon,
But the prettiest thing in the world/Is for me without exaggeration: Hamburg's Köhlbrand Bridge over the
Elbe River."*

Translation of Gunter Gabriel's 1999 song *My Bridge*

Egon Jux (born 1927)

1962–66 designs the 355-meter-long Grand Duchess Charlotte Bridge in Luxembourg
1974 builds the Köhlbrand Bridge in Hamburg
1975 wins the European Steel Prize for the "continent's loveliest bridge" founds "Jux und Partner," an architects' practice in Darmstadt Consults on the construction of the pedestrian bridge over the "Alte Fahrt" in Potsdam

The Berlin-born songwriter and singer Gunter Gabriel throws an admiring glance at Hamburg's Köhlbrand Bridge every time he looks out of his houseboat moored along the banks of the river Elbe. On the occasion of its twenty-fifth anniversary, the singer was inspired to compose the song.

The possibility of building a bridge across the Köhlbrand channel was first discussed in April 1968 when Hamburg's Senate finally rejected a tunnel option. Two years later, Egon Jux, a local architect and pupil of Le Corbusier, started work on the Köhlbrand Bridge. Having created the 355-meter-long Grand Duchess Charlotte Bridge in Luxembourg, he was felt to be the best-qualified man to build a crossing over the channel that would replace a ferry which carried 6,800 vehicles a day and which was prone to the vagaries of the weather. Jux designed a bridge with 3,940 meters of approaches and a span of 520 meters supported by two 135-meter-high steel towers. It would be the fourth-largest cable-stay bridge in Europe with a carriageway made of concrete box segments. Below the box girder on which the carriageway rests, the A-shaped towers bend in the way, thus lending them a graceful appearance despite their

huge size. The arrangement of 88 steel cables was intended to support the 4,840-ton carriageway forever. With overhead clearance of 53 meters, even the largest of ships would be able to sail beneath it.

September 20, 1974 was the big day: the Köhlbrand Bridge was finished and Hamburg's city fathers looked on with pride. The German President Walter Scheel (in office 1974–79) came in a motor launch and enthusiastically led the way across the bridge; 100,000 followed him. The number of vehicles projected to use the bridge daily was 25,000; critics argued that this was grossly overestimated, but they were proved wrong. A year later, Jux was awarded the European Steel Prize for the continent's loveliest bridge.

Three years later, the newspapers started referring to the Köhlbrand Bridge as the "Maria Callas of bridges: attractive, unpredictable and no stranger to scandal," alluding to the fact that its steel cables had to be replaced, having rusted far more than expected — and the job cost millions. The Köhlbrand Bridge may have cost € 82 million thus far, but it still enjoys a reputation as something of a triumphal arch, Hamburg's very own "Golden Gate."

"Hamburg's Golden Gate": Since 1974, the Köhlbrand Bridge has spanned the 300-meter-wide Köhlbrand channel that links the Norder and Süder arms of the river Elbe. It connects the container terminal and the free port

For Expo '92
Seville: Alamillo Bridge 1987–92

"A number of engineers tried to prove that it would not hold up, or that it would be much more costly than expected. They even commissioned someone to redo the calculations."

Santiago Calatrava on the construction
of his Alamillo Bridge, 2001

Santiago Calatrava (born 1951)

1951 born July 28 in Benimamet near Valencia
1969–74 studies architecture at the "Escuela Technica Superior de Arquitectura de Valencia"
1975–81 studies civil engineering at the ETH Zurich; earns PhD
1981 opens own architectural and civil engineering practice in Zurich
1987 awards the Auguste Perret Prize of the "Union Internationale des Architectes"
1987–92 builds the Alamillo Bridge for Expo '92 in Seville
opens his second practice in Paris
awarded "Médaille d'Argent de la Recherche et de la Technique," Paris
1991 awards honorary degree from the University of Seville
European Glulam (glued laminated timber construction) award, Munich
1992 member of the "Real Academia de Bellas artes des San Carlos," Valencia

When architect Santiago Calatrava presented his design for a bridge at Seville's "Expo '92" (April 20–October 12, 1992), it was met with deep suspicion. Its 142-meter-high pylon, set at an angle of 58 degrees to reflect the Great Pyramid at Giza, appeared to be much too daring. Calatrava was inspired by the idea that planning a bridge is an act of far greater cultural significance than building a museum. In his own words: "A bridge is more efficient as it can be used by everyone. A bridge can be accessed by people who are not interested in art. One single gesture transforms nature and gives it order. Nothing is more efficient."

The idea of constructing a slanting pylon came to Calatrava while studying his 1986 sculpture *Running Torso* in which wire holds five marble cubes balanced along a diagonal line. The intention was to capture the tension and the forces in a body moving forward. It was this dynamism, as exemplified by the flight of a heron, that was to be inherent in the Alamillo Bridge. Its mass was to be a sufficient counterweight to the decking. Thirteen pairs of cables would lend additional stability, making braces superfluous. The span would measure 200 meters as it crossed the Meandro San Jerónimo.

Across the Guadalquivir river, 1.5 km away, Calatrava planned to erect a second bridge with a pylon that formed both a mirror image of his Alamillo Bridge and an aesthetic counterpart to it. Both pylons were to act as a monumental gateway to the north of Spain. Calatrava was able to sell the idea of his daring Alamillo Bridge to the regional government of Andalusia, but for financial reasons his proposal for a bridge across the Guadalquivir was not adopted.

Calatrava's additional design for a 526.5-meter-long viaduct across the neighboring peninsula of La Cartuja was given the go-ahead, however, and was to provide access to the exhibition site from the north. Future plans for the viaduct included a 22-meter-wide carriageway to carry traffic heading to Mérida and Portugal on its upper tier and a 4.4-meter-wide walkway and cycle path for pedestrians and cyclists on the lower tier.

The Alamillo Bridge and the Cartuja Viaduct, together with the re-designed railway station and theater, modern hotels and roads, were designed to improve the municipal and regional infrastructure. Calatrava's pylon became something much more significant, however, and is now the symbol of modern Seville.

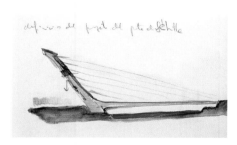

Design for the Alamillo Bridge, Santiago Calatrava

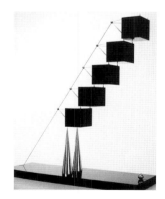

Calatrava's sculpture *Running Torso* was the inspiration for the Alamillo Bridge, Bauschänzli Restaurant, Zurich 1986

The daring cantilever of the 250-meter-long Alamillo Bridge is the symbol of modern Seville

Tempting Fate
Kobe: Akashi Kaikyo Bridge 1988–98

"Pilot ropes, consisting of polyaramid fiber, are crossed over each span by helicopter. Catwalk ropes are erected by means of the hauling system, and floor systems are attached to complete the aerial walkways... After installation of girders, corrosion protection is conducted by means of wire wrapping and rubber wrapping."

Cable work procedures, 1988

Projects initiated by the Honshu-Shikoku Office of Bridge Construction (started 1970)

1979 Ohmishima Bridge
1985 Ohnaruto Suspension Bridge (span of 876 m)
1988 Seto Ohashi Bridge
1988–98 Akashi Kaikyo Bridge
1999 Tatara Bridge: with a central span of 890 meters, this is the largest, double-decked cable-stayed bridge in the world
1999 Kurushima and Kaikyo Bridges with spans of 1,020 and 1,030 meters respectively

These were the instructions that engineers from the Honshu-Shikoku Office of Bridge Construction used in their attempt to build a suspension bridge with the world's longest central span across the Aleshi Strait in Japan. It was said that the engineers chose not to tempt fate by building a bridge of the record length of two kilometers, making do instead with 1,991 meters. On January 17, 1995, the major earthquake that struck the Hanshin metropolitan area raised questions about even that degree of modesty. In the nearby city of Kobe, 5,000 people were killed and 27,000 were injured. Fifty thousand buildings collapsed, leaving 300,000 people homeless and wandering around the ruined city. The world directed its gaze upon the Akashi Kaikyo Bridge that was then still under construction. Had it survived this natural disaster? Did this signal the end of "the building of the Tower of Babel"? Could the Danes with their East Bridge across the Great Belt now claim to have the world's longest suspension bridge? The worst was to be feared as the earthquake had measured 7.2 on the Richter scale. The engineers were relieved to discover that their bridge had held! Admittedly, it had been built to withstand an earthquake measuring 8.5 on the Richter scale.

Serious doubts arose, however, when measurements revealed that the earthquake had caused both 297-meter-towers to move 80 cm apart. Engineers solved the problem by performing new structural analyses of the hangers and the decks and erecting a stabilizing truss. The opening ceremony could thus go ahead as planned on April 5, 1998 when the Japanese crown prince and his consort, representing the Tenno Akihito (crowned emperor in 1989), walked across it accompanied by 1,500 invited guests. The Akashi Shikoku Bridge was the last link in the easternmost of the three multi-bridge crossings between the islands of Honshu and Awaji. Ease of movement among Japan's islands was thus guaranteed beyond the Awaji islands as far as the Ohnaruto suspension bridge. Since then, Japanese car and truck drivers have been able to travel as far as Naruto on Shikoku island. The link that is planned for the future between Honshu and Shikoku on the route from Onomichi to Imabari will eclipse even the dimensions of the Akashi Shikoku project.

Steel caisson used in laying the foundations of the 297-meter-high towers

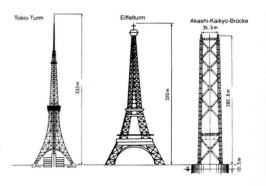

A tower of the Akashi Kaikyo Bridge in relation to the Tokyo and Eiffel Towers

The Akashi Kaikyo Bridge with its central span of 1,991 m and an overall suspended length of 3,911 m has been the world's longest suspension bridge since 1998

World Record for Spinning

Spanning the Great Belt: Denmark's East Bridge 1991–98

"The countries involved in the project all flew flags on the spinning wheels that—to cheers from the delighted workers—were set in motion for the 4,662nd—and last—crossing of the Storebælt, heading towards a world cable-spinning record seven weeks ahead of schedule."

Press release by A/S Storebæltsforbindelsen,
November 20, 1996

When A/S Storebæltsforbindelsen, a state-owned company established in 1987, announced the completion of cable-spinning operations on the East Bridge, a fixed link between the Danish islands of Fyn and Zealand was within reach for the first time.

This mammoth undertaking started on June 23, 1988 with the boring of the East Tunnel between Zealand and the small island of Sprogø. The West Bridge, a 6.6-km-long girder construction of pre-stressed concrete that connects Sprogø and Fyn, was completed in 1993. Two years earlier, a consortium had started work on the project at the heart of Great Belt Link, the construction of the East Bridge. With a free span measuring 1,624 meters, the East Bridge was to unite the islands of Fyn and Zealand. Spinning of the main cables between the 254-meter-high towers had already dragged on for 4 months. A 5.37-mm-thick continuous loop of wire was spun to form the main cables with a diameter of 82.7 cm. Installation of the 48-meter-long hangers that would carry the deck 65 meters above the water could not start until the next summer. The queen of Denmark, Margrethe II, was finally able to open the 18-km-long Great Belt Link to traffic on June 14, 1998. What had until

then been a two-hour ferry journey was now a car journey of 15 minutes. Getting around Denmark, a country with hundreds of islands, was now much easier. This link was the continuation of what had first started with the construction of the bridge across the Little Belt in 1934 and 1971 and the Fehmarn Strait (Germany) in 1963.

Completion of the Great Belt Link did not signal the end to infrastructure improvements in Denmark, however. In summer 2000, the Öresund Bridge (1997–2000) provided the first fixed road link between Copenhagen and the Swedish city of Malmö. It was already under discussion in 1996 when the men working on the East Bridge were vying with their Japanese counterparts for the honor of building the world's longest suspension bridge and achieving a world cable-spinning record. The Akashi Kaikyo Bridge was completed almost at the same time but exceeded the span of the East Bridge by 366 meters.

Prince Joachim of Denmark inspecting the tunnel work

The East Bridge under construction, August 1997

Only Japan's Akashi Kaikyo Bridge is longer: Denmark's East Bridge across the Great Belt

A blade of light across the Thames

London's Millennium Bridge 1996–2000

"When the Millennium Bridge is open and…you talk to some of those four million or more people who will be using it, you'll realize it's a very, very important benefit. You can see the impact it will have in terms of regeneration, bringing the river to life, much needed new life…The bridge will be decorated with sculptures and be lit at night to form a blade of light across the Thames."

Norman Foster's vision for the Millennium Bridge, 1996

Lord Norman Foster (born 1935)

1935 born June 1 in Manchester
1963 member of Team 4 with his wife Wendy Foster and Richard and Sue Rogers
1979–86 designs the Hong Kong and Shanghai Bank, Hong Kong
1982–85 designs the BBC Radio Centre, London
1984 designs the Carré d'Art in Nimes
1989 designs the Millennium Tower in Tokyo
1990 knighted
1991 awards the French Academy's Gold Medal
1992–99 re-designs the Reichstag, Berlin
1996–2003 redevelopment of Wembley Stadium
1997–2002 designs the Music Centre, Gateshead, Tyne and Wear
1999 becomes 21st Pritzker Architecture Prize Laureate

Lord Norman Foster's vision became reality when engineers Ove Arup & Partners completed the design that Foster had developed with the sculptor Sir Antony Caro. The first River Thames crossing built in over 100 years, the Millennium Bridge was opened to the public on June 10, 2000. That first weekend, more than 100,000 Londoners and tourists crossed the foot-bridge that cost in excess of €27.6 million to build. A strikingly modern steel and aluminum construction with the appearance of a single sweeping arc, the bridge's four-meter-wide aluminum deck is supported by two elliptical piers and is carried by four 120-millimeter cables that are anchored in concrete abutments on either side of the river. It offers uninterrupted views of numerous London landmarks along the Thames.

Londoners were now able to take a leisurely stroll from the City and St. Paul's Cathedral to Tate Modern, without the stench and noise of busy traffic. Unfortunately, the freedom was short-lived. The Millennium Bridge had to be closed only eight days later when reports began to appear in the London press about bridge users feeling "seasick." Intensive tests revealed that synchronized footfall caused the bridge to sway by as much as

70 millimeters, a phenomenon not previously encountered in this type of construction. Ninety-one damping mechanisms weighing 700 tons in total were installed below the deck to reduce movement to a millimeter. The damping mechanisms function similarly to shock absorbers, softening the horizontal sway experienced on the bridge. A second type of damping mechanism, added as a precaution, softens vertical motion. The process took two full years, however, not a matter of weeks, as Foster and Partners had assumed, before the "blade of light" was re-opened to the public.

The Millennium Bridge was closed shortly after opening because of structural problems, 1996–2000

Glossary

Abutments

the supports of a bridge on either bank that receive the thrust of an arch

Aerodynamic Stability

the ability of a bridge deck to resist the force of the wind without sustaining damage from torsion or oscillation

Anchorage

massive block of natural stone or concrete in which a suspension bridge's cable ends are fixed

Aqueduct

a bridge or channel carrying water

Arched bridge

a bridge whose deck rests upon one or more semi-circular or segmental arches

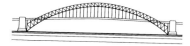

Hell Gate Bridge, New York, 1917

Bascule

a moving bridge with weights at either end which when engaged cause the span to rise

Beam bridge

a bridge using a beam of stone, concrete, wood or another material

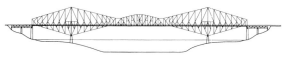

Quebec Bridge, Canada, 1919

Box girder

a beam with a hollow square or rectangular section

Cable

the main cable of a suspension bridge strung between two banks and to which suspenders are affixed

Cable-spinning

technique first used on the Brooklyn Bridge to spin the main cables of a suspension bridge

Cable-stayed bridge

a bridge whose deck is supported by inclined cables affixed to its pylon(s)

Caisson

a chamber kept watertight by means of compressed air; used to build bridge foundations, it sinks as material is removed from the riverbed

Cantilever

a beam supported at one end and unsupported and free at the other

Cantilever bridge

a bridge with arms projecting from piers, usually with a central suspended section

Cellular construction

in American bridge-building, method used to construct towers from relatively small steel box units welded together

Clapper bridge

a prehistoric bridge made of stone slabs

Cutwater

end of a pier base, usually pointed, facing upriver

Extrados

outer surface of the curve of an arch

Fender

a protective structure surrounding a pier construction

Fortified tower

tower defending the entrance to a bridge

Glass fiber
a reinforcing material with high tensile strength

Harp configuration
on a cable-stayed bridge, the parallel cables connecting the pylon and deck

Jack-knife bridge
moving bridge whose deck hinges upwards in the center

Machicolation
opening between the corbels of a projecting parapet of a fortified wall or tower through which stones or burning objects could be dropped on attackers

Oscillation
typically vertical movement of a suspension bridge in the wind

Pier
the support between two bridge spans, often arches

Pontoon bridge
bridge made of boats or other floating units tied together

Pre-stressed concrete
concrete in which stretched (pre-tensioned) steel strands are embedded

Pylon
vertical element to which cable stays are fixed

Rope suspension bridge
a wooden footbridge supported by slack ropes

Reinforced concrete
concrete with embedded steel rods that give greater tensile strength

Segmental arch
an arch formed of a circular arc less than a semi-circle

Shear
force acting across any beam or structural unit

Shear strength
force inherent in diagonal struts that prevents a bridge from moving apart

Spandril
the almost triangular space between one side of the outer curve of an arch, a wall and the ceiling or framework

Stiffening truss
a truss beneath the deck of a suspension bridge

Suspenders (Hangers)
the wires or bars hanging from a suspension bridge's main cables and which carry the deck.

Suspension bridge
a bridge whose deck is supported from above by long ropes, cables or chains hanging from towers or fixed to the shore

Major Suspension Bridges in the World

1991 m: Akashi Kaikyo, Japan 1998
1624 m: Great Belt East, Denmark 1998
1410 m: Humber, U.K. 1981
1385 m: Jangyn, China 1999
1377 m: Tsing Ma, Hong Kong 1999
1298 m: Verrazano Narrows, New York/U.S.A. 1964
1280 m: Golden Gate, California/U.S.A., 1937
1210 m: Hoga Kusten, Sweden, 1997
1158 m: Mackinac, Michigan/U.S.A., 1957
1100 m: Minami (South) Bisan-Seto, Japan 1988

George Washington Bridge, New York, 1931

Suspension types

brace suspension bridge (primitive)

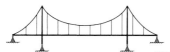
suspension bridge (modern)

anchored suspension bridge

cable-stayed bridge

Swing bridge

a moving bridge whose deck can be swung open

Tension

a force that lengthens members

Torsional strength

prevents a bridge from twisting, for instance by means of struts

Tower

vertical element from which the cables of a suspension bridge are hung

Truss

a framework of members in tension and compression making a long-span beam

Truss frame

frame that is flexible on account of its oblique struts

Vibration dampers

counterweights that reduce bridge vibration

Zig-zag bridge

traditional bridge design in China, with decks arranged at right angles

Bibliography and Suggested Readings

In Exmoor National Park: Tarr Steps pages 12–13

Sources

"Teufelslegende über die Tarr Steps," In:
   http://www.btinternet.com/~exmoor/framegz.htm
Blackmore, Richard Doddridge, *Lorna Doone*, London 1869.

Suggested Readings

Adams, Chris, *Circular walks from Tarr farm at Tarr steps*,
   Walkscene 1999.
*A Quick Guide to Winsford on Exmoor, what to see, where it is,
   and a map, a walk over Windford's eight bridges & notes on
   Winsford Hill & Tarr Steps*, London 1982.
Miller, G.R., Miles J., Heal, O.W., *Moorland Management, A Study
   of Exmoor, A Report commissioned by the Countryside
   Commission*, Cambridge 1984.
Snell, Frederick John, *The Country of the Wild Red Deer, A hand-
   book for Dulverton and the Exmoor border land*, London 1898.
Sudell, Henry Darrell, *Guide to Tarr Steps*, Exmoor, Taunton 1928.

The beam bridges of Afghanistan pages 14–15

Sources

Borcherding, Gisela (Ed.), *Granatapfel und Flügelpferd, Märchen
   aus Afghanistan*, Kassel 1975, pp. 17, 33.
Hackin Ria, Kohzad, Ahmad Ali (Ed.), *Légendes et coutumes
   afghanes*, Paris 1953.
Lebedew, K. (Ed.), *Afganskie skazki*, Moscow 1955.
Parker, Barret, Javid, Ahmad (Ed.), *A Collection of Afghan Legends*,
   Kabul 1970.
Zachová, Eliska (Ed.), *Die Fee aus dem Granatapfel und andere
   afghanische Märchen*, Prague 1961.

Suggested Readings

Aarne, Antti, Thompson, Stith (Ed.), *The Types of the Folktale*,
   FFC 184, Helsinki 1961.
Clifford, Mary L., *The Land and People of Afghanistan*, Philadelphia
   1973.
Jason, Heda, "Types of Jewish-Oriental Oral Tales," *Fabula 7*,
   Berlin 1964/65.

Nepal: Bridges across the Dudh Khosi
and Trisuli Rivers pages 16–17

Sources

Heunemann, Annette (Ed.), *Der Schlangenkönig, Märchen aus
   Nepal*, Kassel 1980.
Kretschmar, Monika (Ed.), *Märchen und Schwänke aus Mustang
   (Nepal)*, pp. 14, 27, 116, 156.
Unbescheid, Günter (Ed.), *Märchen aus Nepal, Die Märchen der
   Weltliteratur, Begründet von Friedrich von der Leyen*,
   Cologne 1987.

Suggested Readings

Moran, Kerry, *Nepal Handbook*, Chico 1991.
Korn, Wolfgang, *The Traditional Architecture of the Kathmandu
   Valley*, Kathmandu 1993.
Shaha, Rishikesh, *Ancient and Medieval Nepal*, New Delhi 1992.

The Aqueduct at Segovia pages 18–19

Sources

Vituvius, Marcus, *Libri decem de architectura*, Ed. Curt Fenster-
   busch, Darmstadt 1981, pp. 393–9.

Suggested Readings

Evans, Harry B., *Aqueduct Hunting in the Seventeenth Century:
   Raffaele Fabretti's De Aquis Et Aquaeductibus veteris Romae*,
   Michigan 2002.
Dupre, Judith, Gehry, Frank O., *Bridges*, New York 1997.

The Alcántara Bridge pages 20–21

Sources

Gajus Julius Lacer, "Grabinschrift," *El Puente de Alcantara en su
   contexto historico*, Ed. Don Antonio Blanco Freijeiro, Don Diego
   Angulo Iniguez, Madrid 1977, p. 38.
Baer, Frank, *Die Brücke von Alcántara*, Rome/Munich 1988.

Suggested Readings

Scarre, Chris, *Chronicle of the Roman Emperors: The Reign-By-
   Reign Record of the Rulers of Imperial Rome*, London 1995.
Glanville, John, Grosant, A.B. (Ed.), *Voyage to Cadiz in 1625*, New
   York 1987.

Rome's Ponte S. Angelo pages 22–23

Sources

Alberti, Leon Battista, *Zehn Bücher über die Baukunst*, Ed.
   Max Theuer, Darmstadt 1991, pp. 436–7.
Bernini, Domenico, *Il Tribunale della S. Rota Romana – totius
   Christiani Orbis supremum Tribunal*, Rome 1717.
Borboni, Andrea, *Delle Statue*, Rome 1661.
Filareti, Antonio di Pietro Averlino, Trattato d'architectura, Lib. XIII,
   In: *Quellenschriften für Kunstgeschichte*, Ed. W. von Oettingen,
   vol. III, Vienna 1890, pp. 386 ff.
Palladio, Andrea, *Die vier Bücher zur Architektur, Nach der Ausgabe
   Venedig 1570 I quattro libri dell'architettura*, Ed. Andreas Beyer,
   Ulrich Schütte, 2nd ed., Zurich/Munich 1984, p. 234.

Suggested Readings

Birley, Anthony R.: Hadrian, *The restless Emperor*, London/
   New York 1997, pp. 283–5.
Boatwright, Mary T., *Hadrian and the Cities of the Roman Empire*,
   Princeton 2000, pp. 162–71.
Weil, Mark S., *The History and Decoration of the Ponte S. Angelo*,
   University Park 1974.

Zhao Xian, Hebei: The An Ji Bridge pages 24–25

Sources

Anji-Brücke, *Kanäle, Brücken und Zisternen, Berichte aus den
   antiken Kulturen Ägyptens, Mesopotamiens, Südwestarabiens,
   Persiens und Chinas*, Berlin 1999, p. 85.

Suggested Readings

Shu-ch'eng, Liang, *A Pictural History of Chinese Architecture,
   A Study of the Development of Its Structural System and the
   Evolution of Its Types*, Ed. Wilma Fairbank, Massachusetts 1984,
   pp. 175–7.
Dupre, Judith, Gehry, Frank O., *Bridges*, New York 1997.

Puente la Reina: the Pilgrims' Bridge   pages 26–27

Sources

Codex Calixtinus, Santiago, *Archivo de la Cathedral*, Ms. 1, fol. 192r.

Herbers, Klaus (Ed.): *Der Jakobsweg, Mit einem mittelalterlichen Pilgerführer unterwegs nach Santiago de Compostela*, 4th ed., Tübingen 1991, p. 86.

Suggested Readings

Meyers, Joan, Simmons, Marc, Pierce, Donna, *Santiago: Saint of Two Worlds*, New Mexico, 1991.

Hall, James A., *A Journey to the West by Domenico Laffi: The Diary of a Seventeenth Century Pilgrim from Bologna to Santiago de Compostela*, New York 1998.

Regensburg's Stone Bridge   pages 28–29

Sources

Monumenta Germaniae Historica, Die Urkunden der deutschen Könige und Kaiser, vol. X, part IV, Hannover 1990, p. 40.

Sachs, Hans, Gesammelte Werke, Ed. A.v. Keller and E. Goetze, vol. 23, Tübingen 1895, p. 326 (Bibliothek des Literarischen Vereins in Stuttgart, vol. 207).

Zeiler, Martin, *Teutsches Reyßbuch durch Hoch- und Nider Teutschland*, Strasbourg 1632, p. 69.

Suggested Readings

Heyman, Jaque, *The Masonry Arch*, New York 1982.

Pearce, Martin, Jobson, Richard, *Bridge Builders*, New York 2002

Avignon: St. Bénézet's Bridge   pages 30–31

Sources

Albanès, J.M., "La légende de Saint Bénézet," 1876, *Villeneuve lez Avignon, Notes historiques*, Ed. Noël Lacombe, Avignon 1990.

Deutschmann, Gerhard, *Sur le pont d'Avignon, Für gemischten Chor*, Augsburg 1969.

Suggested Readings

Chance, Jane, *Medieval Mythography: From the School of Chartres to the Court of Avignon, 1177–1350*, Florida 2000.

Dupre, Judith, Gehry, Frank O., *Bridges*, New York 1997.

Puente de Órbigo   pages 32–33

Sources

Cervantes Saavedra, Miguel de, *Der sinnreiche Junker Don Quijote von der Mancha*, Ed. Ludwig Braunfels, Book 1, Chapter 49, Stuttgart/Hamburg 1966, pp. 510–12, 534.

"En la Historia del Rey Don Iuan el segundo, Se dize lo siguiente al proposito de la historia del passo ponroso," cap. 240, *Libro del passo honroso defendido por el excelente cavallero Suero de Quiñones, Copilado de un libro antiguo de mano por F. Iuan de Pineda*, Salamanca 1588, Reprint: Ed. Martín de Riquer, Madrid 1970, pp. 23–4.

Lena, Pedro Rodríguez de, *Libro del passo honroso defendido por el excelente cavallero Suero de Quiñones*, Puente de Órbigo 1434.

Suggested Readings

Jaâen, Didier Tisdel, *John II of Castile and the grand master Alvaro de Luna: a biography compiled from the chronicles of the reign of King John II of Castile (1405–1454)*, Castalia.

Round, Nicholas, *The Greatest Man Uncrowned: A Study of the fall of Don Alvaro de Luna (Series A: Monografias, No. 1111)*, London 1986.

The Kapell Bridge in Lucerne   pages 34–35

Sources

"Luzerner Ratsbeschluss von 1599," *Die Brücken von Luzern*, Ed. Adolf Reinle, 1963, p. 4.

Starklof, Ludwig, *Tagebuch meiner Wanderung durch die Schweiz*, Bremen/Leipzig 1819, pp. 253–5.

Zelger, Franz (Ed.), *Luzern im Spiegel alter Reiseschilderungen, 1757–1835*, Lucerne 1933.

Suggested Readings

Brown, David J., *Bridges: Three Thousand Years of Defying Nature*, London 1999.

Cahors: The Valentré Bridge   pages 36–37

Sources

Martinot, Robert, Polomski, Joël, *Le Diable du pont Valentré*, Castelnau-Montratier 1990, pp. 1–28.

Martinot, Robert, *Légendaire de Quercy*, 2nd ed., Saint-Céré 1984.

Suggested Readings

www.marie-cahors.fr

www.cambridge2000.com

www.uq.edu.au

www.wadsworth.com

Florence: Ponte Vecchio   pages 38–39

Sources

Vasari, Giorgio, *Leben der ausgezeichnetsten Maler, Bildhauer und Baumeister von Cimabue bis zum Jahre 1567*, with Ludwig Schorn, Ernst Förster, Ed. Julian Kliemann, Worms 1988, p. 285.

Vasari, Giorgio, *Le Vite de' più eccellenti Architetti, Pittori et Scultori Italiani da Cimabue insino a' tempi nostri descritte in lingua Toscana*, Florence 1550.

Suggested Readings

Brucker, Gene, *Florence, The Golden Age, 1138–1737*, Berkeley/Los Angeles/London 1998, p. 13.

Eaton, Ruth, "Florence, Ponte Vecchio," *Living Bridges, The Inhabited Bridge, Past, Present and Future*, Ed. Peter Murray, Mary Anne Stevens, Munich/New York 1996, pp. 62–5.

Fletcher, Banister, *A History of Architecture*, London 1987, p. 757.

Ladis, Andrew, *Taddeo Gaddi, Critical Reappraisal and Catalogue Raisonné*, Columbia/London 1982, pp. 17–80.

Verona: the Scaliger Bridge   pages 40–41

Sources

Erdmannsdorff, Friedrich Wilhelm von, *Kunsthistorisches Journal einer fürstlichen Bildungsreise nach Italien 1765/66*, translated from the French manuscript with an explanatory text, and edited by Ralf-Torsten Speler, Munich/Berlin 2001, pp. 91–3.

Torello, Saraina, *Le histoire, e fatti de' veronesi nei tempi del popolo, e signori scaligeri*, Verona 1549, p. 49.

Suggested Readings

Dupre, Judith, Gehry, Frank O., *Bridges*, New York 1997.

Brown, David J., *Bridges*: *Three Thousand Years of Defying Nature*, London 1999.

Prague: Charles Bridge   pages 42–43

Sources

Jenstein, Johannes von, *Bericht an Papst Bonifaz IX.* (1393), In: Cod. vat. lat. 1122, Bibliotheca Apostolica Vaticana, fol. 162r–169r.

Haselbach, Thomas Ebendorfer von, *Chronica regum et imperatorum Romanorum libri septem*, book 6, Vienna 1449 (Cod. 3423, Österreichische Nationalbibliothek, Handschriften- und Inkunabelsammlung).

Suggested Readings

Brown, David J., *Bridges: Three Thousand Years of Defying Nature*, London 1999.

Bridges: Their Art, Science and Evolution, Random House Value, New York 1983.

The château of Chenonceaux   pages 44–45

Sources

Bullant, Jean, *Reigle generalle d'architecture des cinq manieres de colonnes*, Paris 1568.

Champigneulle, Bernard: *Loire-Schlösser*, 3rd ed., Munich 1971, pp. 194–200.

Nostradamus, Michael, *Les premières centuries ou propheties*, Ed. Macé Bonhomme de 1555, Geneva 1996, book 1, verse 35.

Suggested Readings

Blunt, Anthony, *Philibert Delorme*, London 1958.

Cloulas, Ivan, *Diane de Poitiers*, Paris 1997.

Mostar: "Stari most," the Old Bridge   pages 46–47

Sources

Bajezidagi, "Pascha," *Mostar, Ed. Salih Rajkovi*, Belgrade 1965, p. 11.

Koschnick, Hans, Schneider, Jens, *Brücke über die Neretva, Der Wiederaufbau von Mostar*, Munich 1995, pp. 7–27.

Mui, Omer, *Mostar u turskoj pjesmi iz XVII vijeka*, Prilozi Orijentalnog instituta, Sarajevo 1969, pp. 94–6.

Suggested Readings

Dupre, Judith, Gehry, Frank O., *Bridges*, New York 1997.

Paésiâc, Amir, *The Old Bridge (Stari Most) in Mostar*, Research Centre for Islamic History, Art, and Culture.

Shanghai: The Zig-Zag Bridge   pages 48–49

Sources

Trigault, Nicolas, *De christiana expeditione apud Sinas suscepta ab societate Jesu*, Augsburg 1615.

Suggested Readings

Lanning, G., Couling, Samuel, *The History of Shanghai*, Shanghai 1921.

Dupre, Judith, Gehry, Frank O., *Bridges*, New York 1997.

Bassano del Grappa: Ponte degli Alpini   pages 50–51

Sources

Palladio, Andrea, *Die vier Bücher zur Architektur*, based on the 1570 Venice edition *I quattro libri dell'architetura*, Ed. Andreas Beyer, Ulrich Schütte, 2nd ed., Zurich/Munich, 1984, p. 230.

Palladio, Andrea, *Scritti sull'architettura (1554–1579)*, Ed. Lionello Puppi, Vicenza 1988, p. 165.

Suggested Readings

Boucher, Bruce, *Andrea Palladio, The architect in his time*, New York/London 1998, pp. 187–90.

McQuillen, Th., Pollalis, S.N., *Palladio's Bridge at Bassano, A Structural Analysis Case Study*, Harvard University, Graduate School of Design, Laboratory for Construction Technology, Boston 1989.

Dupre, Judith, Gehry, Frank O., *Bridges*, New York 1997.

Pont Neuf: The New Bridge   pages 52–53

Sources

Mercier, Louis Sébastian, *Tableau de Paris*, Amsterdam, pp. 1782–88.

En vertu des lettres patentes du 16 mars 1578, Voir Archives Nationales, ZIF 1065, Plumitif de la construction du Pont-Neuf.

Suggested Readings

Dupre, Judith, Gehry, Frank O., *Bridges*, New York 1997.

Stack, Edward M., *Le Pont Neuf, a structural review*, 3rd ed., New York.

The Rialto Bridge across the Grand Canal   pages 54–55

Sources

Palladio, Andrea, Die vier Bücher zur Architektur, Nach der Ausgabe Venedig 1570 *I quattro libri dell'architetura*, Ed. Andreas Beyer, Ulrich Schütte, 2nd ed., Zurich/Munich, 1984, p. 239.

Suggested Readings

Brown, David J., *Bridges: Three Thousand Years of Defying Nature*, London 1999.

Dupre, Judith, Gehry, Frank O., *Bridges*, New York 1997.

Isfahan: Allahverdi Khan bridge   pages 56–57

Sources

Morier, James Justinian, *A Journey through Persia, Armenia and Asia Minor to Constantinople in the Years 1808 and 1809*, London 1812.

Morier, James Justinian, *A second Journey through Persia, Armenia and Asia Minor to Constantinople, between the Years 1810 and 1816*, London 1818.

Morier, James Justinian, *Reisen durch Persien in den Jahren 1808 bis 1816*, Ed. Karl Klaus Walther, Wolfgang Barthel, Berlin, 1985, p. 188.

Suggested Readings

Blair, Sheila S., Bloom, Jonathan M., *The Art and Architecture of Islam*, New Haven/London 1994.

Blunt, W., *Isfahan, Pearl of Persia*, London 1974.

Chanby, Sheila R., *The Golden Age of Persian Art 1501–1722*, London 1999.

Hillenbrand, Robert, *Islamic Architecture, Form, Function and Meaning*, Edinburgh 1994.

McChesney, Robert, "Four Sources on Shah Abba's Building of Isfahan," *Muquarnas*, vol. 5 (1988), pp. 103–34.

Michell, George, *Architecture of the Islamic World*, London 1978.

Savory, Roger, *Iran under the Safawids*, Cambridge 1980.

Welch, Anthony, *Shah Abbas and the Arts of Isfahan*, New York 1979.

Venice: The Bridge of Sighs   pages 58–59

Sources

Casanova, Giacomo, *Histoire de ma fuite des prisons de la République de Venise qu'on apelle les Plombs*, Prague 1788.

Casanova, Giacomo, *Meine Flucht aus den Bleikammern von Venedig*, trans. Ulrich Friedrich Müller, Kristan Wachinger, Munich 1998, p. 18.

Grillparzer, Franz, *Tagebuch der Reise nach Italien*, Venice 1819.

Suggested Readings

www.venetia.it

www.republicofvenice.com

Beijing: The Jade Belt Bridge   pages 60–61

Sources

Desroches, Jean Paul, Yuanming Yuan, *Die Welt als Garten, Europa und die Kaiser von China*, Ed. Berliner Festspiele GmbH, Frankfurt am Main 1985, p. 123.

Suggested Readings

Fugl-Meyer, H., *Chinese Bridges*, Shanghai/Hong Kong 1937, pp. 85 ff.

Neville-Hadley, Peter, *China, the Silk Routes*, Old Saybrook 1997, pp. 356–7.

Sirén, Osvald, *The Imperial Palaces of Peking*, 3 vols, Paris/Brussels 1926.

Ssu-ch'eng, Liang, *A Pictorial History of Chinese Architecture, A Study of the Development of Its Structural System and the Evolution of Its Types*, Ed. Wilma Fairbank, Massachusetts 1984, pp. 175–81.

The Iron Bridge, Coalbrookdale   pages 62–63

Sources

Mehrtens, G., *Eisenbrückenbau*, Leipzig 1908, p. 269.

Riedel 1797, vol. I, p. 157, cited from: Fuchtmann, Engelbert, *Stahlbrückenbau, Bogenbrücke, Balkenbrücke, Fachwerkbrücke, Hängebrücke*, 2nd ed., Munich 1996, p. 14.

Suggested Readings

Chaplin, Robin, "New Light on Thomas Farnolls Pritchard," *Shropshire Newsletter*, June 1968.

Harris, John, Pritchard Re-devivus, *Country Life*, vol. 29. February 1968.

Hume, John, "Cast Iron and Bridge-Building in Scotland," *Industrial Archaeology Review*, vol. 11 (1978), pp. 290–9.

Hutton, Charles, Tracts on Mathematical and Philosophical Subjects, Nr. 6: *The History of Iron Bridges* (1812), pp. 146–66.

Miller, S., "The Iron Bridge at Sunderland, A Revision," *Industrial Archaeology Review*, vol. 11 (1976), pp. 70–2.

Smith, J., *A conjectural Account of the Erection of the Iron Bridge*, London 1979.

Trinder, Barrie, *The Industrial Revolution in Shropshire*, Chichester 1973, pp. 5–47.

Trinder, Barrie, "The First Iron Bridges," *Structural Iron, 1750–1850*, Ed. R.J.M. Sutherland, Alderhot/Brookfield/Singapore/Sydney 1997, pp. 248–56 (Studies in the History of Civil Engineering, Ed. Joyce Brown, Vol. 9).

Bristol: The Clifton Suspension Bridge   pages 64–65

Sources

Brunel, Isambard Kingdom, Letter to his politican brother-in-law Benjamin Hawes, Bristol 1829.

Keefer, Samuel, Report of Samuel Keefer, civil engineer, to the president and directors of the Niagara Falls, and to the president and directors of the Clifton Suspension Bridge Company, Ontario 1869.

Webb, W.W., *A complete account of the origin and progress of the Clifton Suspension Bridge over the River Avon*, Bristol 1864.

Wright & Sons (Ed.), *History and Construction of the Clifton Suspension Bridge: illustrated by view of the bridge and sections of the various parts*, Bristol 1864.

Suggested Readings

Body, Geoffrey, *Clifton Suspension Bridge, An illustrated history*, Bradford-on-Avon 1976.

Gomme, Andor, Jenner, Michael, Little, Bryan, *Bristol, an architectural history*, London 1979, pp. 335–48.

Jenkins, David, Jenkins, Hugh, *Isambard Kingdom Brunel, engineer extraordinary*, Hove 1977.

Kentley, Eric, Hudson, Angie, Peto, James, *Isambard Kingdom Brunel, recent works*, London 2000.

McIlwain, John, *Clifton Suspension Bridge*, Andover 1996.

Meynell, Laurence Walter, *Isambard Kingdom Brunel*, London 1955.

Meynell, Laurence Walter, *Builder and Dreamer, A life of Isambard Kingdom Brunel*, London 1957.

Noble, C.B, *The Brunels*, London 1938.

Pugsley, Alfred, *The Works of Isambard Kingdom Brunel, An engineering appreciation*, London 1976.

Rolt, Lionel Thomas Caswall, *Isambard Kingdom Brunel*, Harmondsworth 1985.

Tames, Richard, *Isambard Kingdom Brunel, An illustrated life of Isambard Kingdom Brunel, 1806–1859*, Princes Risborough 2000.

Williamson, Peter A. *Isambard Kingdom Brunel and the building of the Clifton Suspension Bridge, A dramatised scientific celebration*, Langport 1985.

Budapest: the Széchenyi Chain Bridge   pages 66–67

Sources

Clark, Adam, "Brief an seine Eltern am 27. Mai 1849," *Vigh, Annamária: The Successes and Anxieties of an Expatriate, Extracts from Adam Clark's letters to his parents, 1834–49, The Széchenyi Chain Bridge and Adam Clark*, Budapest 1999, p. 130.

Clark, Adam, *Briefe an seine Eltern (1834–58)*, Historisches Museum, Budapest, Inventarisationsbücher Nr. VII und VIII.

Clark, Adam, *Einige Worte über den Bau der Ofner-Pesther Kettenbrücke*, Pest 1843.

Suggested Readings

Barany, G., *Stephen Széchenyi and the Awakening of Hungarian Nationalism, 1791–1841*, Princeton 1968.

Bibó, István, "The Chain Bridge as Architectural Creation," *The Széchenyi Chain Bridge and Adam Clark*, Budapest 1999, pp. 167–75.

Brody, Judit, "William Tierney Clark: Civil Engineer," *The Széchenyi Chain Bridge and Adam Clark*, Budapest 1999, pp. 60–77.

Gall, Imre, "Chain Bridge – the History of its Construction," *The Széchenyi Chain Bridge and Adam Clark*, Budapest 1999, pp. 131–66.

Holló, Szilvia Andrea, "Széchenyi's Role from the Idea to the Passing of the Bridge Act," *The Széchenyi Chain Bridge and Adam Clark*, Budapest 1999, pp. 12–44.

Parsons, Nicholas T., "Romaticism, Nationalism and Reform in Adam Clark's Scotland and Count István Széchenyi's Hungary," *The Széchenyi Chain Bridge and Adam Clark*, Budapest 1999, pp. 45–59.

Smith, D., "The works of William Tierney Clark (1783–1852), Civil Engineer of Hammersmith," *Newcomen Society Transactions*, vol. 63 (1991/92), pp. 181–207.

The Göltzsch Valley Bridge in the Vogtland pages 68–69

Sources

Schubert, Johann Andreas, "Die Überbrückung des Göltzsch- und des Elstertales auf der sächsisch-bayerischen Eisenbahn," *Allgemeine Bauzeitung*, vol. 11, Vienna 1846, pp. 382–92.

Schubert, Johann Andreas, *Theorie der Construction steinerner Bogenbrücken*, Dresden/Leipzig 1847/48.

Schubert, Johann Andreas, "Offene Erklärung 27.4.1848," *Dresdner Journal*.

Suggested Readings

www.user.tu-chemnitz.de

James Eads' St. Louis Bridge pages 70–71

Sources

Eads, James Buchanan, *Report of the Chief Engineer*, St. Louis 1870–1871.

Eads, James Buchanan, *Addresses and Papers*, Ed. Estill McHenry, St. Louis 1884.

Warren, Major G.K., "Report on Bridging the Mississippi River," *Annual Report of the Chief of Engineers for 1878*, Washington 1878, pp. 1058–60.

Suggested Readings

Bryan, William H., "The Engineers' Club of St. Louis, Its History and Work," *Journal of the Association of Engineering Societies*, St. Louis 1900, pp. 161–2.

Dorsey, Florence, *Road to the Sea, The Story of James B. Eads and the Mississippi River*, New York 1947.

Gies, Joseph, *Bridges and Men*, Garden City, N.Y. 1963.

Kouwenhoven, John A., "The designing of the Eads Bridge," *Structural Iron and Steel 1850–1900*, Ed. Robert Thorne, Aldershot - Burlington/Singapore/Sydney 2000, pp. 161–94 (Studies in the History of Civil Engineering, vol. 10).

McCabe, James D., "James B. Eads," *Great Fortunes and How They Were Made*, Philadelphia/New York/Boston 1871, pp. 211–12.

Miller, Howard S., *The Eads Bridge*, Columbia/London 1979.

Steinman, David B., *The Builders of the Bridge*, New York 1945, pp. 170–3.

Volti, Rudi, *Caisson, The Facts on file Encyclopedia of Science, Technology and Society*, New York 1999.

Woodward, Calvin M., *A History of the St. Louis Bridge, Containing a Full Account of Every Step in Its Construction and Erection and Including the Theory of the Ribbed Arch and the Tests of Materials*, St. Louis 1881.

New York: Brooklyn Bridge pages 72–73

Sources

Crane, Hart, *Complete poems*, Ed. Brom Weber, Newcastle upon Tyne 1987.

Miller, Henry, *Reise nach New York*, Ed. Marlis Pörtner, Zurich 1962.

Pope, Thomas, *A Treatise on Bridge Architecture, in which the Superior Advantages of the Flying Pendant are fully proved*, New York 1811.

Roebling, John, "Bridging the East River," *Architects' and Mechanics' Journal*, vol. 1, New York, March 31, 1860, p. 209.

Roebling, John, "Correspondence," *Architects' and Mechanics' Journal*, vol. 2, New York, June 14, 1860, pp. 13–14.

Schuyler, Montgomery, "The Bridge as a Monument," *Harper's Weekly 27*, New York, May 26, 1883, p. 326.

Suggested Readings

Billington, David P., *The Tower and the Bridge, The New Art of Structural Engineering*, New York 1983, pp. 72–87.

Haw, Richard, *Brooklyn Bridge, The Ideology of the Opening Day*, Odense 2001.

Howells, W.D., *Impressions and Experiences*, New York 1896, p. 227.

McCullough, David, "The Treasure from the Carpentry Shop, The Extraordinary Original Drawings of the Brooklyn Bridge," *American Heritage 31*, New York, December 1979, pp. 19–29.

Nevins, Deborah et al., *The Great East River Bridge, 1883–1983*, Brooklyn 1983.

Reier, Sheron, *The Bridges in New York*, New York 1977, pp. 10–27.

Schuyler, Montgomery, "Art in Modern Bridges," *Century Magazine*, vol. 38, New York, May 1900, pp. 12–25.

Schuyler, Montgomery, *The Roeblings, A Century of Engineers, Bridgebuilders and Industrialists*, Princeton 1931.

Snyder, Robert, *This is Henry, Henry Miller from Brooklyn*, Los Angeles 1974.

Snyder-Grenier, Ellen M., *Brooklyn! An Illustrated History*, Philadelphia 1996, pp. 67–81.

Steinman, David B., *The Builders of the Bridge, The Story of John Roebling and his Son*, New York 1945.

Steinman, David B., Watson, Sarah Ruth, *Bridges and their Builders*, New York 1957, pp. 205–47.

Porto's Maria Pia Bridge and Luís I Bridge pages 74–75

Sources

Seyrig, Théophile, *Le pont du Douro à Porto*, Paris 1878.

Eiffel, Gustave, *Biographie industrielle et scientifique*, Paris 1920.

Suggested Readings

Harriss, Joseph, *The Tallest Tower, Eiffel and the Belle Epoque*, Boston 1975, pp. 43–5, 232–3.

Dupre, Judith, Gehry, Frank O., *Bridges*, New York 1997.

The Forth Bridge pages 76–77

Sources

Baker, Benjamin, *The Forth Bridge*, London 1882.

Fontane, Theodor, *Sämtliche Werke*, vol. 20, Munich 1962, p. 165.

Rowand, Anderson R., *Transactions of the National Association for the Advancement of Art and its Application to Industry*, Edinburgh Meeting 1889, London 1890, p. 153.

Westhofen, Wilhelm, "The Forth Bridge," *Engineering Magazine*, London 1890, Reprint: Edinburgh 1989.

Suggested Readings

Hamond, Rolt, *The Forth Bridge and its Builders*, London 1964.

Koerte, Arnold, *Two Railway Bridges of an Era, Firth of Forth and Firth of Tay, Technical Progress, Disaster and New Beginning in Victorian Engineering*, Basel/Boston 1991.

Macdonald, Angus J., "The Technology of the Forth Bridge," *John Fowler, Benjamin Baker, Forth Bridge*, Ed. Iain Boyd Whyte, Angus H. Macdonald, Stuttgart/London 1997, pp. 12–19.

Mackay, Sheila, *The Forth Bridge, A Picture History*, Edinburgh 1990.

Morris, William, *News from Nowhere and Selected Writings and Designs*, Harmondsworth 1984, p. 332.

Murray, Anthony, *The Forth Railway Bridge, A Celebration*, Mainstream, Edinburgh 1983.

Paxton, Roland, *100 Years of the Forth Bridge*, London 1990.

Phillips, Philip, *The Forth Bridge in its Various Stages of Construction and Compares with the Most Notable Bridges of the World*, Edingburgh 1893.

Whyte, Iain Boyd, "A Sensation of immense Power," John Fowler, Benjamin Baker, *Forth Bridge*, Ed. Iain Boyd Whyte, Angus H. Macdonald, Stuttgart/London 1997, pp. 6–11.

London: Tower Bridge   pages 78–79

Sources

Ceremonial to be observed at the Opening of the Tower Bridge, London 1894.

Goble, W., Austin, H., *The Tower Bridge*, London 1894.

"The Tower Bridge: its history and construction from the date of the earliest project to the present time," Engineer Office, pp. 106, London 1894.

The Tower Bridge, *The Builder*, June 30, 1894.

Webb, Edward Brainerd, Bolland, James, *London Bridge, Shall London Bridge be widened, or shall a new bridge be built near the Tower?*, London 1877.

Suggested Readings

Billington, David P., Perkins, David, *The Tower and the Bridge, The new art of structural engineering*, New York 1983.

Godfrey, Honor, *Tower Bridge*, London 1988.

Rose, Alan, *Build your own Tower Bridge and Tower of London*, London 1982.

Tonge, Elizabeth, *Bridge by bridge through London the Thames from Tower Bridge to Teddington*, Whitstable 1989.

Williams, Archibald, *The romance of modern engineering, containing interesting descriptions in non-technical language of the Nile Dam, the Panama Canal, the Tower Bridge, the Brooklyn Bridge, the Trans-Siberian Railway, the Niagara Falls Power Co., Bermuda Floating Dock, etc.*, London 1903.

Wilson, Ken, *The story of Tower Bridge*, London 1985.

Dresden: The "Blue Wonder"   pages 80–81

Sources

Köpcke, Claus, *Über die Beseitigung der Schwankungen an einer Hängebrücke*, Dresden 1885.

Köpcke, Claus, *Über die Schwingungen eiserner Brücken*, Dresden 1995.

Weyern, Willy von, "Bei Blasewitz schwebt eine Hängebrücke," *Blaues Wunder, Dresdens wunderlichste Brücke*, Ed. Michael Wüstefeld, Berlin/Brandenburg 2002, p. 34.

Suggested Readings

www.wh2.tu-dresden.de

www.virtual-dresden.de

Paris: Alexander III Bridge   pages 82–83

Sources

Heredia, José-Maria de, "Salut à l'Empereur," *Ponts de Paris, Architecture et histoire*, Ed. Jocelyne Deputte, Paris 1994, p. 178.

Résal et Alby, *Notes sur la construction du pont Alexandre III*, Paris 1899.

Schneider & Cie, "Le Pont Alexandre III à Paris," *Extrait des nouvelles annales de la constuction*, 1899.

Suggested Readings

www.pariswater.com

Spirn, Michele Sober, *The Bridges of Paris–Going To Series: Going To Paris*, 2000.

Quebec City: Quebec Bridge   pages 84–85

Sources

Cooper, Theodore, *American Railroad Bridges*, New York 1898.

Cooper, Theodore, *General Specifications for Steel Railroad Bridges and Viaducts*, New York 1901.

Cooper, Theodore, *General Specifications for Foundations and Substructures of Highway and Electric Railway Bridges*, New York 1902.

Dawson, S.E. (ed.), *Quebec Bridge Inquiry Report, Royal Commission on Collapse of Quebec Bridge*, Ottawa 1908.

Suggested Readings

Innis, Hug R., Bilingualism and Biculturalism, *An Abridged Version of the Royal Commission Report*, Toronto 1973.

Middleton, William, D., *The Bridge at Québec*, Bloomington 2001.

Proposed Bridge over the St. Lawrence at Quebec, 1885, reprint: London 1981.

The Quebec Bridge over the St. Lawrence River near the City of Quebec on the Line of the Canadian National Railway, Report, Ottawa 1919.

Schneider, C.C., *Report and Plans, Also Report on Design of Quebec Bridge*, Ottawa 1908.

Sydney Harbour Bridge   pages 86–87

Sources

Bradfield, John Job Carew, *Materials from the professional library of John Job Carew Bradfield relating to engineering railways, bridges etc., also included are materials on the Sydney Harbour Bridge*, Sydney 1915–1932.

Bradfield, John Job Carew, *Contract for the construction of a cantilever bridge across Sydney River Harbour from Dawes Point to Milson's Point, Sydney, New South Wales, Australia, specification*, Sydney 1921.

Bradfield, John Job Carew, *The City and Suburban Electric Railways and the Sydney Harbour Bridge*, Sydney 1924.

Bradfield, John Job Carew, *Sydney Harbour Bridge and City Railway*, Willoughby 1932.

Suggested Readings

Billington, Robert and Sarah, *The Bridge*, Mount Kuring-Gai 1999.

Building a bridge for Sydney, The North Sydney Connection, Ed. North Sydney Council Historical Services Dept., North Sydney 2000.

Colett, Betty V., Lee, Philip, McDougall, Reece, *Balain to the City foreshores study Iron Cove Bridge to Sydney Harbour Bridge*, Ed. S.S. Clark, North Ryde 1979 (Environmental and urban studies report, Nr. 52).

Environmental Impact Statement for Ninth Lane and Footway on Sydney Harbour Bridge, Ed. GHD Transportation Consultants Pty Ltd, Sydney 1982.

Hill, Deirdre, *A Bridge of Dreams*, Sydney 1982.

Kent, Alfred James, *The Bridge opened*, Illustrated Supplement to Official Souvenir, March 1932, Sydney 1932.

Sands, John, Keast, Burke, *Achievement, A Collection of unusual Studies of the Sydney Harbour Bridge*, Sydney 1932.

Sydney Harbour Bridge Conservation Management Plan, Ed. Heritage Group, Department of Public Works and Services, Sydney 1998.
Walker and Kerr, *National Trust Classification Card – Sydney Harbour Bridge*, Sydney 1974.

New York: George Washington Bridge   pages 88–89

Sources

Ammann, Othmar Hermann, *Study of a Highway Bridge across the Hudson River at New York between Washington Heights and Fort Lee*, New York 1923, 1925.
Ammann, Othmar Hermann, *The Problems of Bridging the Hudson River at New York*, Procedures of the Connecticut Society of Civil Engineers, New York 1924.
Ammann, Othmar Hermann, *The Hudson River Bridge at New York between Fort Washington and Fort Lee*, Procedures of the Connecticut Society of Civil Engineers, New York 1928.
Ammann, Othmar Hermann, *The Cables of the Hudson River Bridge*, Published by the Port of New York Authority, New York 1929–1930.
Ammann, Othmar Hermann, The Port of New York Authority, *Progress Reports on Hudson River Bridge at New York between Fort Washington and Fort Lee*, New York 1929–1931.
Ammann, Othmar Hermann, *George Washington Bridge, General Conception and Development of Design, Transactions of the American Society of Civil Engineers*, vol. 97, New York 1933.
Le Corbusier, "A Place of Radiant Grace," *When the Cathedrals were White*, New York 1947.

Suggested Readings

Asplund, S.O., *On the Deflection Theory of Suspension Bridges*, Stockholm 1943.
Kunz, F.C., *Design of Steel Bridges, Theory and Practice for the use of Civil Engineers and Students*, New York 1915.
Rastorfer, Darl, *Six Bridges: The Legacy of Othmar H. Ammann*, New York 2000, pp. 39–76.
Volse, Louis A., "Othmar Hermann Ammann: An Artist in Steel Design," *Engineering News-Record*, 15. May 1958.
Wadell, I.A.L., *Bridge Engineering*, New York 1923.

San Francisco: Golden Gate Bridge   pages 90–91

Sources

John Bernard McGloin, S.J., "Symphonies in Steel: San Francisco Bay Bridge and the Golden Gate," *San Francisco, The Story of a City*, San Rafael, California 1978.
Morrow, Ivring Foster, *Monastery of the Visitation of the Blessed Virgin Mary*, San Francisco 1919.
O'Shaughnessy, Michael M., *Hetch Hetchy, Its Origin and History*, San Francisco 1934.
Strauss, Joseph Baermann, *The Golden Gate Bridge at San Francisco, California, Report of the Chief Engineer with Architectural Studies and Results of the Fact-Finding Investigation*, San Francisco 1930.
Strauss, Joseph Baermann, "The Mighty Task is Done," *San Francisco News*, San Francisco, May 26, 1937.
Strauss, Joseph Baermann, "Golden Gate Bridge," *San Francisco Chronicle*, San Francisco, May 27, 1937.
Strauss, Joseph Baermann, *The Golden Gate Bridge, The Chief Engineer's Final Report*, San Francisco 1938.

Suggested Readings

Adams, Charles F., *Heroes of the Golden Gate*, Palo Alto 1987, pp. 325–38.
Brown, Allen, *Golden Gate, Biography of a Bridge*, Garden City 1965.
Cassady, Stephen, *Spanning the Gate*, Mill Valley 1979.
Dillon, Richard, Moulin, Thomas, DeNevi, Don, *High Steel, Building the Bridges across San Francisco Bay*, Millbrae 1979.
Van der Zee, John, *The Gate, The True Story of the Design and Construction of the Golden Gate Bridge*, New York 1986.

The Kwai Bridge   pages 92–93

Sources

Boulle, Pierre, *Le pont de la rivière Kwai*, Ed. René Julliard, 1952.
Boulle, Pierre, *The Source of the River Kwai*, London 1967.
Boulle, Pierre, *Die Brücke am Kwai*, novel, Gottfried Beutel, Erich Thanner, Vienna/Hamburg 1988, pp. 161, 250.

Suggested Readings

Anderegg, Michael A., *David Lean*, Boston 1984.
Brownlow, Kevin, David Lean, *A biography*, New York 1996, pp. 345–59.
Pratley, Gerald, *The Cinema of David Lean*, New York 1974.
Silver, Alain, Ursini, James, *David Lean and his Films*, Hollywood 1991.

Steward, John, *To the River Kwai*, London 1988.
Dupre, Judith, Gehry, Frank O., *Bridges*, New York 1997.

Innsbruck: The "Europabrücke"   pages 94–95

Sources

Inscription in the chapel on the Europabrücke.

Suggested Readings

Leonhardt, Fritz, *Brücken, Bridges*, Stuttgart 1994, pp. 192–3.
www.bridgeweb.com

New York: Verrazano Narrows Bridge   pages 96–97

Sources

Ammann, Othmar Hermann, "Neue Brücken and Expresstrassen," *Schweizerische Bauzeitung*, 22. February 1958.
Ammann, Othmar Hermann, "The Narrows Bridge at New York," *Procedures of the Connecticut Society of Civil Engineers*, 1960.
Ammann, Othmar Hermann, *Planning and Design of the Verrazano-Narrows Bridge*, Transactions of New York Adademy of Sciences, New York 1963.
Ammann, Othmar Hermann, *The Verrazano-Narrows Bridge, International Harbour Conference*, Antwerp, 1964.
Ammann, Othmar Hermann, "Verrazano-Bridge, History, Financing, Design and Construction, Proceedings of the American Society C.E.," *Journal of the Construction Division 92/CO 2*, March 1966.
Tarrow, Susan (trans.), "Cellere Codex," *The Voyages of Giovanni da Verrazano*, 15241528, Ed. Lawrence C. Wroth, Yale 1970, pp. 133–43.

Suggested Readings

Cohen, Edward, *Notes on the Professional Career ot Othmar Hermann Ammann*, The New York Academy of Sciences, New York 1967.
Fioravante, Janice, "The Farms Gave Way Just 25 Years Ago," *New York Times*, May 18, 1997.
Kempner, Mary Jean, "The Greatest Bridge of Them All", *Harper's Magazine*, November 1964, pp. 70–6.
Mumford, Lewis, "The New York Skyline," *New Yorker*, 14 November 1959, pp. 186–91.
Rasthofer, Darl, *Six Bridges: The Legacy of Othmar H. Ammann*, New York 2000, pp. 135–62.
Tales, Gay, *The Bridge*, New York 1964, Chapter 6.

Panama: Bridge of the Americas/Puente de las Américas   pages 98–99

Sources

Roy, Alonso, "Inauguración del Puente de las Américas," Escritos Históricos de Panamá, 2002. http://www.alonsoroy.com/aroy/cp7.html.

Suggested Readings

Friar, William, *Portrait of the Panama Canal: From Construction to the Twenty-First Century*, Portland 1999.
Major, John, *Prize Possession: The United States and the Panama Canal 1903–1979*, Cambridge 1994.
http://www.structurae.de/de/structures/data/str00448.html.

Lisbon: Ponte de 25 Abril   pages 100–101

Sources

Steinman, David Barnard, *Fifty years of progress in bridge engineering*, New York 1929.
Steinman, David Barnard, *Bridges*, New York 1947.
Steinman, David Barnard, *Engineering report on Mackinac Straits Bridge to Mackinac Bridge Authority*, 1953.
Steinman, David Barnard, *Bridges and their builders*, New York 1957.
Steinman, David Barnard, *Famous Bridges of the World*, New York 1961.

Suggested Readings

Ratigan, William, *Highways over broad waters, Life and times of David B. Steinman, bridgebuilder*, Grand Rapids 1959.

Istanbul: Bosporus Bridge (Bogaziçi)   pages 102–103

Sources

Hochtief-A.G. für Hoch- and Tiefbauten, Issue 47, Essen 1974.
Leonardo da Vinci, Ms. L, fol. 66r, *Bibliothèque de l'Institut de France*, Paris, 1498–1502.

Suggested Readings

Dupre, Judith, Gehry, Frank O., *Bridges*, New York 1997.
Reti, Ladislao (Ed.), *The Manuscripts of Leonardo da Vinci at the Bibl. Nacional of Madrid*, vol. 5, New York 1974.

Hamburg: The Köhlbrand Bridge   pages 104–105

Sources

Gabriel, Gunter, "Die Welt hat viele Wunder," *Hamburger Abendblatt*, September 24, 1999.

Suggested Readings

Brown, David J., *Bridges: Three Thousand Years of Defying Nature*, London 1999.

Seville: Alamillo Bridge   pages 106–107

Sources

Calatrava, Santiago, *Calatrava*, Ed. Philip Jodidio, Cologne 2001, pp. 25–35.
Calatrava, Santiago, *Dynamische Gleichgewichte, Neue Projekte*, Ed. Lyall Sutherland, 3rd ed., Zurich 1993, Foreword.

Suggested Readings

Frampton, Kenneth, Webster, Anthony C., Tischhauser, Anthony, *Calatrava Bridges*, 2nd ed., Basel/Boston/Berlin 1996.
McQuaid, M., *Santiago Calatrava, Structure and Expression*, exhibition catalog, Museum of Modern Art, New York 1993, pp. 38–9.
Sharp, D., *Santiago Calatrava*, London 1992, pp. 40–3.
Dupre, Judith, Gehry, Frank O., *Bridges*, New York 1997.

Kobe: Akashi Kaikyo Bridge   pages 108–109

Sources

Bühler, Dirk, *Brückenbau*, Munich 2000, p. 145.

Suggested Readings

Brown, David J., *Bridges: Three Thousand Years of Defying Nature*, London 1999.
Dupre, Judith, Gehry, Frank O., *Bridges*, New York 1997.

Spanning the Great Belt: Denmark's East Bridge   pages 110–111

Sources

"Presseerklärung des Bauherrn A/S Storebæltsforbindelsen vom 20. November 1996," *Brücken*, Ed. Judith Dupre, Munich 2000, p. 112.

Suggested Readings

Brown, David J., *Bridges: Three Thousand Years of Defying Nature*, London 1999.

London's Millennium Bridge   pages 112–113

Sources

"North and South London linked by new Millennium bridge," *NetLondon*, London, April 28, 1999.
Foster, Norman, *Architecture is about People*, Munich/London/New York 2002.

Suggested Readings

Foster, Norman, Grey, Spencer de, Nelson, David et al., *Foster Catalogue 2001*, Munich/London/New York 2001, pp. 180–3.
Jenkins, David, *On Foster ... Foster On*, Munich/London/New York 2000.
Jodidio, Philip, *Sir Norman Foster*, 2001.
Pawley, Martin, *Norman Foster, A global architecture*, New York 1999.
Rem, Koolhaas, Foster, Norman, Mendini, Alessandro (Ed.), *Colours*, Basel 2001.

Index

127

Front cover and p. 2: Golden Gate Bridge, San Francisco, see p. 90. Inset from left to right: Tower Bridge, London, see p. 78; Alamillo Bridge, Seville, see p. 106, Rialto Bridge, Venice, see p. 54; Akashi Kaikyo Bridge, Kobe, see p. 108; The Forth Bridge, Scotland, see p. 76
Back cover: Charles Bridge, Prague, see p. 42; Allahverdi Khan Bridge, Isfahan, see p. 56; Brooklyn Bridge, New York, see p. 72
Frontispiece: Jade Belt Bridge, Peking, see p. 60
pp. 114/115: Detail of the Akashi Kaikyo Bridge, see p. 108

We would like to thank Dirk Bühler of the Deutsche Museum in Munich for the use of the images on p. 116 and p. 117 (from Bühler, Dirk, *Brückenbau*, Deutsches Museum, Munich 2000)

Die Deutsche Bibliothek – CIP-Einheitsaufnahme data is available
Library of Congress Control Number: 200 210 95 55

© Prestel Verlag, Munich · Berlin · London · New York 2002

Prestel Verlag
Königinstrasse 9, 80539 Munich
Tel. +49 (89) 38 17 09-0 · Fax +49 (89) 38 17 09-35

4 Bloomsbury Place, London WC1A 2QA
Tel. +44 (20) 7323–5004 · Fax +44 (20) 7636–8004

175 Fifth Avenue, New York NY 10010
Tel. +1 (212) 995–2720 · Fax +1 (212) 995–2733
www.prestel.com

Edited by Gabriele Ebbecke

Translated from the German by Stephen Telfer, Edinburgh
Copyedited by James Young, Munich
Designed and Typeset by Meike Weber and Carolin Beck
Lithography by Eurocrom 4, Villorba (TV)
Printed by Sellier, Freising
Bound by Conzella, Pfarrkirchen

Printed in Germany on acid-free paper
ISBN 3-7913-2701-1

Photo Credits

Key: t = top, b = bottom, m = middle, r = right, l = left

AKG p. 44 t. r. and m. r.; Erich Lessing p. 46 m. l.
allOver, Hans-Joachim Doll p. 104 t. l.; Rainer Grosskopf p. 34 t. l., p. 59, p. 86 t. l. and r.; Manfred Hölscher p. 92 t. l, p. 93; Ferdinand Hollweck p. 23, p. 35, p. 53, p. 83; JBE Photo p. 78 t. l. and Front Cover 1 from left, p. 79; Barbara Kirchhof p. 74 t. l.; Rolf E. Kunz p. 19, p. 41, p. 110 t. l.; Govert Vetten p. 45
Georg Anderhub, Luzern p. 34 b. r.
Anzenberger, Gérard Sioen p. 97
architekturphoto p. 89
artur p. 18 t. l., p. 64 t. l., p. 65, p. 68 t. l., p. 69; Achim Bednorz, Monheim p. 37; Klaus Frahm p. 106 t. l. and Front Cover 2 from left, p. 107; Hammel, van der Voort, Monheim p. 42 t. r.; Lisa Hammel, Monheim p. 55; Dieter Leistner p. 67; Florian Monheim p. 44 t. l.; Tomas Riehle p. 46 t. l., p. 76 t. l. and Front Cover 1 from right, p. 77; Wolfgang Schwager p. 66 t. l.
The Bridgeman Art Library p. 56 m. r. and b., p. 62 t. l.
Dirk Bühler, München p. 108 t. l. and Front Cover 2 from right., p. 114/115
Contec-Art, 1.945.018.23 p. 109
Robert Cortright p. 24 t.r., p. 25
Photo Deutsches Museum, Munich p. 12 b. r., p. 28 b. r., p. 68 m. l. and t. r., p. 70 t. r., p. 72 b. l., p. 82 b.
Focus, Udo Tschimel p. 16 b. l.
Focus, Urs F. Kluyver, p. 48 t. l.
Foster and Partners, Nigel Young p. 112 t. l., p. 113
Bernhard Graf p. 26 t. r., p. 28 b. l., p. 32 t. l., p. 33, p. 36 t. l., p. 50 t. l., p. 51
Kulturhistorisches Bildarchiv Hansmann, Munich p. 64 r.
Robert Harding p. 12 t. l., p. 13, p. 17, p. 31, p. 63, p. 103; 2000 Neale Clark p. 72 t. l. and Back Cover; Philip Craven F.P.G., Paul van Riel dez. ´92 p. 49; Surrey Studios p. 30 t. l.; Robert Frerck p. 100 t. l.; Geoff Renner p. 88 t. l.; Ellen Roowey p. 43; Peter Scholey p. 2, p. 61
Matthias Hauer, Munich p. 34 t. r.
Honshu-Shikoku Bridge Authority p. 108 t.r. and b.r.
Bildagentur Huber p. 96 t. l.; Geiersperger p. 94 t. l.; R. Schmid p. 81
IFA-Bilderteam, Arakaki p. 21
The Image Bank, Michael Melford p. 88 r.; 1995 Marvin E. Newman p. 96 t. r.; Guido A. Rossi p. 102 t. l.
Jux-Gruppe p. 105
Architektur-Bilderservice Kandula, M. Krause p. 56 t. l. and Back Cover, p. 57
Kaufmann Grafikdesign p. 26 t. l.
Markus Kirchgessner, Frankfurt p. 98 t. l., p. 99
laif, Gonzalez p. 27; Hoffmann p. 73
LOOK, Karl Johaentges p. 1 and p. 90 t. l.; Christian Heeb Front Cover; Holger Leue p. 91; Konrad Wothe p. 80 t. l.
Daniel Munoz p.85
Claus Pihl p. 111
Andreas Post, Münster p. 22 r., p. 82 t. l.
Walter Raunig, Munich p. 15
Ingo Röhrbein, Hamburg p. 104 t.
David Shogren, St. Louis Missouri p. 71
SLUB, Dt. Fotothek, Martin Würker p. 80 m. r.
Storebault-forbindelsen p. 110 t.r.
Stretto di Messina P. P. A., p. 11
Martin Thomas, Aachen p. 60 b. r.
Transglobe, Rudolf Bauer p. 84 t. l.; Ivo K. Petrik p. 95; P. Spierenburg p. 60 t. l.
Ullstein Bild p. 74 m. l.; Camera Press Ltd. p. 90 m. l.
Marlies Vujovic, Wien p. 47
Thomas Peter Widmann, Regensburg p. 28 t. l., p. 29, p. 42 t. l. and Back Cover, p. 75, p. 87, p. 101
Visum p. 39
ZEFA, Rossenbach p. 58 t. l.
ZUMA Press, J. B. Forbes p. 70 t. l.